MODERN CERTIFICATE MATHEMATICS

MODERN CERTIFICATE MATHEMATICS

Volume 2

COORDINATE GEOMETRY
TRIGONOMETRY
CALCULUS

BY
F. CAPANNI M.A., B.Sc.
Head of the Mathematics Department
Holyrood Secondary School, Glasgow.

Designed by **ROGER SADLER D.A.(Edin.)**

HOLMES McDOUGALL, EDINBURGH

First published 1978
Copyright © 1978 F. Capanni.

Published by Holmes McDougall Ltd.,
            Allander House,
            137–141 Leith Walk,
            Edinburgh, EH6 8NS.

All rights reserved. Except for the purposes of review, fair dealing, or private study, no part of this book may be reproduced in any form or by any means without the prior permission of the publishers.

Set in "Monophoto" Times New Roman 327 and
                   Times Mathematics 569B.

Printed in Great Britain by Holmes McDougall Ltd., Edinburgh.

ISBN: 0 7157 **1357-4**

CONTENTS

COORDINATE GEOMETRY

Unit 1: Vectors   9
Components of a vector; distance formula; coordinates of the mid-point; loci.

Unit 2: Straight Lines   19
Gradient of a straight line; equations of straight lines $y = mx$, $y = mx+c$, $ax+by+c = 0$; lines parallel to the axes; perpendicular lines.

Unit 3: The Circle   35
The circle $x^2+y^2 = r^2$; $(x-a)^2+(y-b)^2 = r^2$; the equation $x^2+y^2+2gx+2fy+c = 0$; intersection of line and circle; tangents.

TRIGONOMETRY

Unit 1: Revision   53
Trigonometrical functions for angles from 0° to 360°; functions for complementary, supplementary and related angles; trigonometrical relations; graphs of sine, cosine and tangent functions; polar coordinates; sine and cosine rules; area of triangles.

Unit 2: Circular Measure   81
Trigonometrical functions in radians; rotation of the plane about the origin; addition formulae; formulae for double angles; equations.

Unit 3: The Products and Sums of Sines and Cosines   101
Equations for the sums and products of sines and cosines.

Unit 4: The Function $a \cos x + b \sin x$ 111
The function in degrees and radians; solution of equations.

CALCULUS

Unit 1: The Differential Calculus   125
Notation for functions, gradients, secants and tangents; gradient of the tangent to a curve; shorthand method of finding a derivative, rules for differentiation, application of derivatives to tangents.

Unit 2: Stationary, Turning and Inflexion Points   139
Increasing and decreasing functions; stationary and turning points and values, points of inflexion.

Unit 3: Graphs   147
Practical applications of stationary values; sketching the graphs of certain functions; maxima and minima.

Unit 4: Rates of Change   157
Constant rate of change; changing rate of change; variation with time.

Unit 5: The Integral Calculus   163
Integration as anti-differentiation.

Unit 6: Definite Integrals and Applications   169
The definite integral; the definite integral as an area; area between curves; the volume of a solid of revolution.

Unit 7: Trigonometrical Functions   183
Derivatives of sine and cosine functions; further use of the chain rule; integrals of trigonometric functions.

ANSWERS

Coordinate Geometry   194

Trigonometry   200

Calculus   205

PREFACE

This book originated from a series of notes which I issued to my certificate pupils each year. They formed the nucleus of the work required for the Scottish Certificate of Education Examination at the higher grade. Since most pupils do not wade through large chunks of mathematics in order to find an explanation of some particular difficulty I have kept explanations as concise as possible but have reinforced them with many worked examples. It should be noted, however, that all the proofs required for the S.C.E. examination are included in the text.

Volume 1 contains two sections, algebra and geometry, and Volume 2 three sections, coordinate geometry, trigonometry, and calculus. Each section is divided into several units. As each part of each unit is completed an assignment on that particular part has been inserted and to meet modern examination trends an extra assignment has been included at the end of each unit containing 12 objective items, these items being of three different types. Thus the two volumes contain over 300 objective type questions. Over and above this there are supplementary examples at the end of each section.

These two books cover the syllabus required for the Scottish Certificate Examination at the higher grade. They may be found useful either as standard text or as supplementary material for the ordinary and advanced levels of the various Associated Examining Boards or Joint Matriculation Boards of the English Universities and Schools.

I wish to thank my wife Anne for her help in checking part of the work in the algebra section, Mr. Iain McLean of Holmes McDougall for also checking part of the work and for his patience with my many mistakes in the original draft and lastly Mr. Rodger Sadler for the excellence of the art-work.

Holyrood Secondary School, F.C.
Glasgow,
May, 1977.

NOTATION

Different writers use different letters to denote various sets of numbers. This does not matter, provided the notation is stated clearly at some point in the work and not contradicted later in the same work.

In this book the following notation for sets of numbers is used.

N the set of natural numbers or positive integers $\{1, 2, 3, \ldots\}$.

W the set of whole numbers $\{0, 1, 2, 3, \ldots\}$.

Z the set of integers $\{\ldots -2, -1, 0, 1, 2, \ldots\}$.

Q the set of rational numbers which includes Z together with the positive and negative fractions.

R the set of real numbers which includes numbers such as $\sqrt{2}, -\sqrt{3}, \pi, e, \ldots$.

Hence $N \subset W \subset Z \subset Q \subset R$.

The following are also used:

R^+ the set of positive real numbers which does not include zero.

$\varnothing$ the empty set.

E the universal set.

COORDINATE GEOMETRY

UNIT 1: VECTORS

COMPONENTS OF A VECTOR 1.1

Let $P(x_1, y_1)$ and $Q(x_2, y_2)$ be any two points. In component form the vector represented by $\vec{OP}$ has components x_1 and y_1 and is written as a column vector $\begin{pmatrix} x_1 \\ y_1 \end{pmatrix}$

Since $\vec{PQ} = \vec{OQ} - \vec{OP}$, then $\vec{PQ}$ represents

$$\begin{pmatrix} x_2 \\ y_2 \end{pmatrix} - \begin{pmatrix} x_1 \\ y_1 \end{pmatrix} = \begin{pmatrix} x_2 - x_1 \\ y_2 - y_1 \end{pmatrix}$$

and hence the vector represented by $\vec{PQ}$ has components $(x_2 - x_1)$ and $(y_2 - y_1)$.

1.2 THE DISTANCE FORMULA

If P is the point (x_1, y_1) and Q the point (x_2, y_2) then the length of the directed line segment $\vec{PQ}$, denoted by $|\vec{PQ}|$ is equal to $\sqrt{[(x_2-x_1)^2+(y_2-y_1)^2]}$.

Proof: (see figure 1)

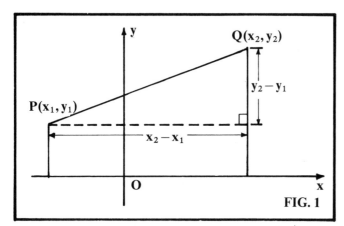

FIG. 1

Since P is the point (x_1, y_1) then $\vec{OP}$ represents $\begin{pmatrix} x_1 \\ y_1 \end{pmatrix}$.

Similarly $\vec{OQ}$ represents $\begin{pmatrix} x_2 \\ y_2 \end{pmatrix}$.

Hence $\vec{PQ} = \vec{OQ} - \vec{OP}$ represents

$$\begin{pmatrix} x_2 \\ y_2 \end{pmatrix} - \begin{pmatrix} x_1 \\ y_1 \end{pmatrix} = \begin{pmatrix} x_2-x_1 \\ y_2-y_1 \end{pmatrix},$$

i.e. the vector represented by $\vec{PQ}$ has components (x_2-x_1) and (y_2-y_1).

$$\Rightarrow PQ^2 = (x_2-x_1)^2+(y_2-y_1)^2$$

$\Rightarrow$ length of the directed line segment $\vec{PQ}$ is

$$|\vec{PQ}| = \sqrt{[(x_2-x_1)^2+(y_2-y_1)^2]}$$

This formula gives the length of the line segment joining the points with coordinates $P(x_1, y_1)$ and $Q(x_2, y_2)$ and is often referred to as the **Distance Formula**.

Example 1

Calculate the length of the line segment joining the points $P(-4, -5)$ and $Q(6, -4)$.

Note: It is better to work with the formula in the form $PQ^2 = (x_2-x_1)^2+(y_2-y_1)^2$ and then take the square root at the end of the calculation.

Hence

$$PQ^2 = (x_2-x_1)^2+(y_2-y_1)^2 = (6+4)^2+(-4+5)^2$$
$$= 100+1 = 101$$

and $PQ = \sqrt{101}$ units.

Example 2

Show that the triangle joining the points with coordinates $A(3, -2)$, $B(8, 3)$ and $C(-3, 4)$ is right angled.

By the distance formula,

$$AB^2 = (x_2-x_1)^2+(y_2-y_1)^2 = (8-3)^2+(3+2)^2$$
$$= 5^2+5^2 = 50$$
$$BC^2 = (x_2-x_1)^2+(y_2-y_1)^2 = (-3-8)^2+(4-3)^2$$
$$= 121+1 = 122$$
$$CA^2 = (x_2-x_1)^2+(y_2-y_1)^2 = (3+3)^2+(-2-4)^2$$
$$= 36+36 = 72$$

Hence $AB^2 + CA^2 = 50+72 = 122$
and $BC^2 = 122$

i.e. $AB^2 + CA^2 = BC^2 \Rightarrow$ triangle ABC is right-angled at A by the converse of Pythagoras.

THE COORDINATES OF THE MID-POINT M OF THE LINE JOINING POINTS $P(x_1, y_1)$ AND $Q(x_2, y_2)$

$P(x_1, y_1)$ and $Q(x_2, y_2) \Rightarrow \vec{PQ}$ represents the vector

$$\begin{pmatrix} x_2 - x_1 \\ y_2 - y_1 \end{pmatrix}$$

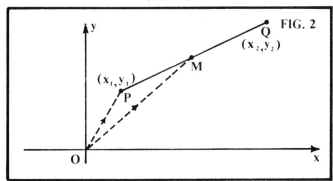

FIG. 2

Since M is the mid-point of PQ then $\vec{PM} = \frac{1}{2}\vec{PQ}$ represents the vector

$$\frac{1}{2}\begin{pmatrix} x_2 - x_1 \\ y_2 - y_1 \end{pmatrix}$$

also $\vec{OP}$ represents the vector

$$\begin{pmatrix} x_1 \\ y_1 \end{pmatrix}$$

so $\vec{OM} = \vec{OP} + \vec{PM}$ represents

$$\begin{pmatrix} x_1 \\ y_1 \end{pmatrix} + \frac{1}{2}\begin{pmatrix} x_2 - x_1 \\ y_2 - y_1 \end{pmatrix}$$

$$= \begin{pmatrix} x_1 + \tfrac{1}{2}x_2 - \tfrac{1}{2}x_1 \\ y_1 + \tfrac{1}{2}y_2 - \tfrac{1}{2}y_1 \end{pmatrix} = \frac{1}{2}\begin{pmatrix} x_2 + x_1 \\ y_2 + y_1 \end{pmatrix}$$

$\Rightarrow$ M has coordinates $\left(\dfrac{x_2 + x_1}{2}, \dfrac{y_2 + y_1}{2}\right)$

Example 3

ABC is a triangle with coordinates A(5, 4), B(6, −8) and C(−4, −5). Find the coordinates of M the mid-point of AC and N the mid-point of AB and show that MN = $\frac{1}{2}$BC.

A(5, 4), C(−4, −5) $\Rightarrow$ M has coordinates

$$\left(\frac{5-4}{2}, \frac{4-5}{2}\right)$$

i.e. $(\tfrac{1}{2}, -\tfrac{1}{2})$

A(5, 4), B(6, −8) $\Rightarrow$ N has coordinates

$$\left(\frac{5+6}{2}, \frac{4-8}{2}\right)$$

i.e. $(\tfrac{11}{2}, -2)$

Also

$$MN^2 = (x_2 - x_1)^2 + (y_2 - y_1)^2 = (\tfrac{1}{2} - \tfrac{11}{2})^2 + (-\tfrac{1}{2} + 2)^2$$
$$= \tfrac{100}{4} + \tfrac{9}{4} = \tfrac{109}{4}$$
$$\Rightarrow MN = \frac{\sqrt{109}}{2}$$

$$BC^2 = (x_2 - x_1)^2 + (y_2 - y_1)^2 = (6+4)^2 + (-8+5)^2$$
$$= 100 + 9 = 109$$
$$\Rightarrow BC = \sqrt{109}$$

hence $MN = \tfrac{1}{2}BC$

Example 4

The vertices of a triangle are A(1, 5), B(4, −1) and C(−2, 4). Find the coordinates of P the mid-point of BC and the coordinates of G which divides AP in the ratio 2 : 1.

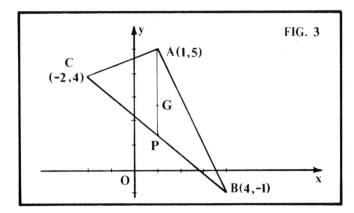

FIG. 3

P has coordinates $\left(\dfrac{-2+4}{2}, \dfrac{4-1}{2}\right)$, i.e. $(1, \tfrac{3}{2})$.

If G divides AP in the ratio $2:1$, then

$$\dfrac{AG}{GP} = \dfrac{2}{1} \Rightarrow \dfrac{x_G - x_A}{x_P - x_G} = \dfrac{2}{1} \quad \text{and} \quad \dfrac{y_G - y_A}{y_P - y_G} = \dfrac{2}{1}$$

$$\Rightarrow \dfrac{x_G - 1}{1 - x_G} = \dfrac{2}{1} \quad \text{and} \quad \dfrac{y_G - 5}{\tfrac{3}{2} - y_G} = \dfrac{2}{1}$$

$$\Rightarrow x_G - 1 = 2 - 2x_G \quad \text{and} \quad y_G - 5 = 3 - 2y_G$$

$$\Rightarrow 3x_G = 3 \quad \text{and} \quad 3y_G = 8$$

$$\Rightarrow x_G = 1 \quad \text{and} \quad y_G = \tfrac{8}{3}$$

i.e. G has coordinates $(1, \tfrac{8}{3})$.

Note: G is called the **centroid** of the triangle, i.e. the point of intersection of the medians.

ASSIGNMENT 1.1

1. Calculate the distance between the following pairs of points. (Answers may be left in surd form.)
    (i) $(5, 6)$ and $(9, 9)$ (ii) $(-4, -3)$ and $(6, -2)$
    (iii) $(-8, 4)$ and $(4, -1)$ (iv) $(7, -2)$ and $(-4, -9)$
    (v) $(3, -5)$ and $(6, 10)$ (vi) $(0, -5)$ and $(8, 0)$
    (vii) $(2ak, 7ak)$ and $(-ak, 3ak)$

2. Show that the triangle joining the points $A(-2, 5)$, $B(6, 2)$, $C(-5, -3)$ is isosceles.

3. The distance between the points $(a, -2)$ and $(3, 2)$ is 5 units. Find the value of a.

4. A line PQ joins the points $P(2, -3)$ and $Q(10, b)$. If PQ has length 10 units find the value of b.

5. Find the coordinates of the mid-point of the straight line joining the following pairs of points.
    (i) $(2, -3)$ and $(4, 8)$ (ii) $(-6, 9)$ and $(-2, -10)$
    (iii) $(6, -5)$ and $(-6, 4)$ (iv) $(2, 0)$ and $(-10, -12)$

6. P, Q, R are the mid-points of the sides of the triangle joining the points $A(2, -3)$, $B(10, 3)$ and $C(9, 4)$. Find the coordinates of P, Q and R and show that triangle PQR is right angled.

7. $A(-1, 5)$, $B(-6, -2)$ and $C(4, 0)$ are the vertices of a triangle. Find the coordinates of M the mid-point of BC and hence the coordinates of G the centroid of the triangle ABC. (*Note:* G divides AM in the ratio $2:1$.)

8. The points $A(6, 4)$, $B(2, 1)$, $C(-6, -5)$ are collinear. Find the ratio $\dfrac{AB}{BC}$.

9. The line joining the points $(-8, 10)$ and $(10, -8)$ is divided into 4 equal parts. Find the coordinates of the points of section.

10. P is the point $(-2, 2)$ and Q the point $(3, 7)$. K lies on PQ and divides PQ in the ratio $3:2$. Find the coordinates of K.

11. Find the lengths of the sides of the parallelogram joining the points A(5, 3), B(−2, 1), C(−4, −3) and D(3, −1). Find also the coordinates of the point of intersection of its diagonals.

12. The mid-point M of the line PQ has coordinates (−1, 3). If P has coordinates (−5, −3) find by calculation the coordinates of Q.

13. A is the point (8, 9), B(−4, 6) and C(10, 3). P, Q and R are the mid-points of AB, AC and BC respectively. Prove that APRQ is a parallelogram and find the lengths of its diagonals.

14. A is the point (1, −5), B(−3, 3), C(5, −3) and D(1, 7). M and N are the mid-points of AB and CD respectively.

    (i) Find the coordinates of M and N.
    (ii) Show that MN = ½BC.
    (iii) Prove that AD bisects BC.

LOCI

The **locus** is the place or region where points are found which satisfy some given condition. The locus is therefore a set of points which may be finite or infinite.

Example 1

What is the locus of points which are 2 cm from the fixed point (3, 2)?

This is the same as saying *Where are the points which are 2 cm from the fixed point (3, 2)?*

Answer: On the circumference of a circle, centre (3, 2) and radius 2 cm.

Example 2

What is the locus defined by

$$\{P(x, y): x \geqq 0, 0 \leqq y < 4, x, y \in R\}?$$

i.e. Where is the set of points with x-coordinate $\geqq 0$ and y-coordinate $\geqq 0$, but < 4?

The locus (or set of points) is shown by the shaded region. (see figure 4)

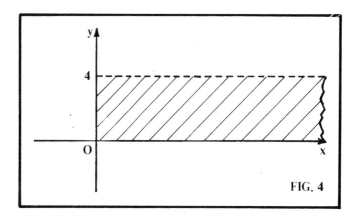

FIG. 4

Note: The boundary line at $y = 4$ is broken to show that the points on the line $y = 4$ are not members of the locus.

We can define locus another way, namely as **the path traced out by a point which moves according to some given condition**.

Example 3

A bullet or shell (regarded as a point) fired from a gun inclined at a certain angle describes a curve called a parabola.

That is, the locus of the bullet or shell is a parabola, the given conditions being the angle and velocity of projection.

Example 4

A point $P(x, y)$ moves so that it is always 3 cm from the x-axis. What is the locus of P?

The locus will be two straight lines each parallel to the x-axis. (see figure 5)

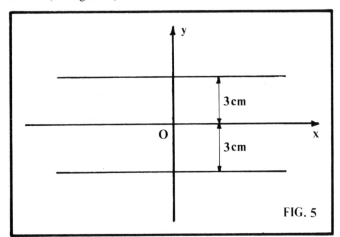

FIG. 5

In coordinate geometry a point $P(x, y)$ may assume any position in the coordinate plane. If $P(x, y)$ changes position according to some geometrical law so that $P(x, y)$ takes up all possible positions governed by the law then there will be an algebraic relation between x and y which is called the **equation of the locus**.

Example 5

A is the point $(2, -2)$ and $P(x, y)$ is a member of the set of points which are the same distance from A as they are from the y-axis. Find the locus of P.

Note: This means we have to find all the possible positions of P satisfying the given condition, which will be some curve in the coordinate plane and which will be defined by the algebraic equation connecting x and y.

Figure 6 shows one position of P at equal distances from A and from the y-axis, where $PM = x$.

Condition is that $PM = PA$
$\Rightarrow \quad PM^2 = PA^2$
$\Rightarrow \quad x^2 = (x-2)^2 + (y+2)^2$, using the distance formula
$\Rightarrow \quad x^2 = x^2 - 4x + 4 + y^2 + 4y + 4$
$\Rightarrow \quad 0 = y^2 - 4x + 4y + 8$

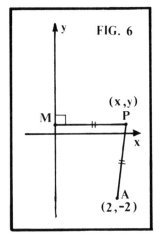

FIG. 6

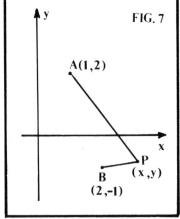

FIG. 7

hence all the possible positions of P lie on the curve with equation $y^2 = 4x - 4y - 8$, which is the equation of the locus of P, and the locus of P is the curve represented by this equation.

Example 6

The distance of a point P from the point $(1, 2)$ is three times its distance from the point $(2, -1)$. Find the equation of the locus of P.

In figure 7 let A be the point $(1, 2)$ and B the point $(2, -1)$. The sketch shows one position of $P(x, y)$ such that $PA = 3PB$.

$$PA = 3PB$$
$$\Rightarrow PA^2 = 9PB^2$$
$$\Rightarrow (x-1)^2 + (y-2)^2 = 9[(x-2)^2 + (y+1)^2],$$
using the distance formula
$$\Rightarrow x^2 - 2x + 1 + y^2 - 4y + 4 = 9[x^2 - 4x + 4 + y^2 + 2y + 1]$$
$$\Rightarrow x^2 + y^2 - 2x - 4y + 5 = 9[x^2 + y^2 - 4x + 2y + 5]$$
$$\Rightarrow 8x^2 + 8y^2 - 34x + 22y + 40 = 0$$
$$\Rightarrow 4x^2 + 4y^2 - 17x + 11y + 20 = 0,$$
which is the equation of the locus of P.

Example 6 may be written down in set notation thus: Since $P(x, y)$ belongs to the set of points which satisfy the condition $PA = 3PB$, then

$$\{P : PA = 3PB\}$$
$$= \{P : PA^2 = 9PB^2\}$$
$$= \{(x, y) : (x-1)^2 + (y-2)^2 = 9[(x-2)^2 + (y+1)^2]\}$$
$$= \{(x, y) : x^2 - 2x + 1 + y^2 - 4y + 4$$
$$= 9(x^2 - 4x + 4 + y^2 + 2y + 1)\}$$
$$= \{(x, y) : 8x^2 + 8y^2 - 34x + 22y + 40 = 0\}$$
$$= \{(x, y) : 4x^2 + 4y^2 - 17x + 11y + 20 = 0\}$$

hence the locus of P has equation
$$4x^2 + 4y^2 - 17x + 11y + 20 = 0$$

Note: When the equation of the locus has been found,
(i) all points whose coordinates satisfy the equation of the locus will be members of the set of points which satisfy the given condition,
(ii) all points which are members of the set of points satisfying the given condition will satisfy the equation of the locus.

ASSIGNMENT 1.2

1. State in words or in set notation the locus of the following.
   (i) The centre of a car wheel of diameter 50 cm.
   (ii) The bob of a swinging pendulum.
   (iii) A man ascending in a lift in a set of high flats.
   (iv) A child sliding down a chute.
   (v) The set of points 5 cm from the point $(0,0)$.
   (vi) The set of points such that $x = 3, y = -1$.
   (vii) The set of points such that $x = \pm 4$.
   (viii) The set $S = \{(x, y) : x > 0, x, y \in R\}$.
   (ix) The set $P = \{(x, y) : x > 0, y < 0, x, y \in R\}$.

2. Show in a diagram the locus (i.e. the set of points) defined by,
   (i) $\{(x, y) : x = y, \ x, y \in R\}$
   (ii) $\{(x, y) : y = -2x, \ y, x \in R\}$
   (iii) $\{(x, y) : -1 \leq x \leq 3, x, y \in R\}$
   (iv) $\{(x, y) : -4 < x < -2, 0 \leq y \leq 3, x, y \in R\}$

3. Sketch on the same diagram the following loci,
   $$P = \{(x, y) : -1 \leq x \leq 4, x, y \in R\}$$
   and $\quad Q = \{(x, y) : -2 \leq y \leq 3, \ x, y \in R\}.$
   Write down in set notation $P \cap Q$.

4. List the set of points given by,
   $$\{(x, y) : y = 2x, -6 \leq x \leq 2, -4 \leq y \leq 4, x, y \in Z\}$$

5. Find the points which are equidistant from the points $A(1, -1)$ and $B(3, 5)$. (This is the same as saying *State the locus of points which are equidistant from the points $A(1, -1)$ and $B(3, 5)$.)*

6. A is the point $(2, -3)$ and $B(3, -2)$. A variable point $P(x, y)$ is such that $PA = PB$. Find the equation of the locus of P.

7. Points are situated at equal distances from the fixed point $(4, 2)$. State the locus of these points.

8. A is the point $(3, -1)$ and $P(x, y)$ is a point whose distance from the point $(3, -1)$ is twice its distance from the x-axis. If M is the perpendicular distance from P to the x-axis, write down the relation between PA and PM. Use this relation to find the equation of the locus of P.

9. Find the equation of the locus of a point P which is equidistant from the points $(2, 0)$ and $(0, -4)$. Sketch the locus.

10. P is a member of the set of points whose distance from $A(4, -2)$ is twice the distance from $B(3, 2)$. Find the equation of the locus of these points.
    Which if any of the following points lie on the locus?
    (i) $(4, 2)$ (ii) $(-1, 3)$ (iii) $(0, \tfrac{8}{3})$ (iv) $(0, 4)$

11. A is the point $(2, 0)$ and B is $(8, 0)$. Find the equation of the loci defined by the following conditions, where $P(x, y)$ belongs to the loci. (Answers may be left in set notation.)
    (i) $\{P: PA = PB\}$ (ii) $\{P: PB = 2\}$
    (iii) $\{P: PA = \tfrac{1}{2}PB\}$ (iv) $\{P: PA = 3PB\}$
    Describe geometrically the loci given by (i) and (ii).

12. $P(x, y)$ belongs to the locus given by
$$\{P: PA^2 - PB^2 = 12\}$$
where A is the point $(0, 2)$ and B is $(0, 8)$. Find the equation of the locus of P and describe it geometrically.
Which if any of the following points lie on the locus?
(i) $(6, 0)$ (ii) $(1, -6)$ (iii) $(2, 3)$
(iv) $(a, 6)$ (v) $(6, k)$

Objective test items

In the objective test items situated at the end of each unit, answer the items as follows (unless otherwise stated).

(i) In questions 1–8 the correct answer is given by one of the options A, B, C, D or E.

(ii) In questions 9 and 10 one or more of the three statements is/are correct.
Answer A if statement (1) only is correct.
       B if statement (2) only is correct.
       C if statement (3) only is correct.
       D if statements (1), (2) and (3) are correct.
       E if some other combination of the given statements is correct.

(iii) In questions 11 and 12 two statements are numbered (1) and (2).
Answer A if (1) implies (2) but (2) does not imply (1).
       B if (2) implies (1) but (1) does not imply (2).
       C if (1) implies (2) and (2) implies (1).
       D if (1) denies (2) or (2) denies (1).
       E if none of the above relationships hold.

ASSIGNMENT 1.3

Objective type items testing sections 1.1–1.4.
Instructions for answering these items are given on page 16.

1. If P is the point $(-8, 4)$ and $Q(-7, -5)$ then $\vec{PQ}$ when written in component form represents the vector,

    A. $\begin{pmatrix} 1 \\ -9 \end{pmatrix}$

    B. $\begin{pmatrix} -1 \\ -1 \end{pmatrix}$

    C. $\begin{pmatrix} 1 \\ -1 \end{pmatrix}$

    D. $\begin{pmatrix} -15 \\ -1 \end{pmatrix}$

    E. $\begin{pmatrix} -1 \\ 9 \end{pmatrix}$

2. The length of the line segment joining the points $(-8, -9)$ and $(-4, -1)$ is

    A. $2\sqrt{31}$
    B. $2\sqrt{37}$
    C. $4\sqrt{5}$
    D. $4\sqrt{10}$
    E. $4\sqrt{13}$

3. If P is the point $(-4, -7)$ and the mid-point M of PQ has coordinates $(2, -5)$, then Q has coordinates

    A. $(-6, -2)$
    B. $(2, -5)$
    C. $(\frac{1}{2}, -\frac{7}{2})$
    D. $(-\frac{7}{2}, \frac{1}{2})$
    E. $(8, -3)$

4. X is the point $(-2, 3)$ and Y the point $(4, -7)$ and $P(x, y)$ belongs to the locus given by $\{P: PX = PY\}$. The equation of the locus of P is

    A. $3x - 5y + 13 = 0$
    B. $10x - 6y + 39 = 0$
    C. $3x + 5y - 13 = 0$
    D. $3x - 5y - 13 = 0$
    E. $10x - 6y - 39 = 0$

5. P is the point $(-2, 4)$, $Q(4, 1)$, $R(1, -5)$ and $S(-5, -2)$. PQRS is a

    A. kite
    B. rhombus
    C. rectangle
    D. square
    E. some other quadrilateral

6. P is the point $(-4, 0)$, Q is $(0, 3)$ and R is $(-1, 1)$. The fourth vertex S of the parallelogram PQRS has co-ordinates,

    A. $(-3, 2)$
    B. $(-5, -2)$
    C. $(3, -2)$
    D. $(5, 2)$
    E. some other set of coordinates

7. The locus of a point $P(x, y)$ is defined by the equation $x^2 + y^2 + 6x - 8y - 15 = 0$. Which one of the following points lies on the locus of P?

    A. $(5, 5)$
    B. $(-5, -2)$
    C. $(0, 0)$
    D. $(5, 2)$
    E. none of these points

8. P is the point $(\cos \alpha, \sin \alpha)$ and Q the point $(\cos \beta, \sin \beta)$. The length of the line segment PQ is given by

   A. $2 \sin \tfrac{1}{2}(\alpha - \beta)$
   B. $\sqrt{2} \cos (\alpha - \beta)$
   C. $\sqrt{2} \sin \tfrac{1}{2}(\alpha - \beta)$
   D. $2 \cos (\alpha - \beta)$
   E. $2 \cos \tfrac{1}{2}(\alpha + \beta)$

9. P, Q and R are the points $(3, -1)$, $(1, 7)$ and $(-7, 5)$.

   (1) triangle PQR is right-angled.
   (2) triangle PQR is isosceles.
   (3) triangle PQR has area 34 units^2.

10. A rod PQ of length 16 cm has its ends P and Q on two rectangular axes OX and OY with P at the point $(\alpha, 0)$ and Q at the point $(0, \beta)$.

    (1) $\alpha^2 + \beta^2 = 256$.
    (2) $x = \dfrac{\alpha}{2}$ and $y = \dfrac{\beta}{2}$ are the coordinates of M the mid-point of PQ.
    (3) $x^2 + y^2 = 64$ is the equation of the locus of M if the rod starts to slide.

11. (1) The point $Q(p, q)$ lies on the locus of $P(x, y)$ defined by $xy = c^2$.
    (2) $pq = c^2$.

12. (1) $PQ^2 = (a_1 - a_2)^2 + (b_1 - b_2)^2$.
    (2) P is the point (a_1, b_1) and Q is the point (a_2, b_2).

UNIT 2: STRAIGHT LINES

THE GRADIENT FORMULA

In figure 9, A is the point (x_1, y_1) and B the point (x_2, y_2).

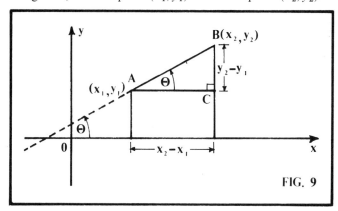

FIG. 9

THE GRADIENT OF A STRAIGHT LINE

The **gradient** of a straight line is the tangent of the angle the line makes with the positive direction of the x-axis. It is denoted by m.

Gradient of AB $= m_{AB} = \tan \theta$.

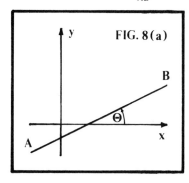

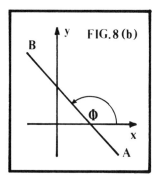

Note that the gradient of a line can be positive or negative.

In figure 8(b), $m_{CD} = \tan \phi$, where ϕ is the obtuse angle and the tangent of an obtuse angle is negative.

Then
$$m_{AB} = \tan \theta = \frac{CB}{AC}$$
$$= \frac{y_2 - y_1}{x_2 - x_1}$$

where $x_2 \neq x_1$

i.e. the gradient of the line AB $= m_{AB} = \dfrac{y_2 - y_1}{x_2 - x_1}$

where $x_2 \neq x_1$. Hence any line joining the points (x_1, y_1) and (x_2, y_2) has gradient given by the formula,

$$\frac{y_2 - y_1}{x_2 - x_1}$$

where $x_2 \neq x_1$. This is called the **Gradient Formula**.

Note:

(i) If $x_2 = x_1$ then AB is parallel to the *y*-axis and the formula has no meaning. We therefore say that the gradient of a line parallel to the *y*-axis is undefined.

(ii) If two lines have the same gradient they are parallel. Conversely, parallel lines have the same gradient, i.e. two lines of gradient m_1 and m_2 are parallel $\Leftrightarrow m_1 = m_2$.

Example 1

Find the gradient of the straight lines joining the following pairs of points:

(i) $A(3, -5)$, $B(7, -2)$ (ii) $P(3, 0)$, $Q(-2, 0)$
(iii) $R(0, -4)$, $S(0, 9)$

(i) $A(3, -5)$, $B(7, -2)$

$$\Rightarrow m_{AB} = \frac{y_2 - y_1}{x_2 - x_1} = \frac{-5 + 2}{3 - 7} = \frac{-3}{-4} = \frac{3}{4}$$

(ii) $P(3, 0)$, $Q(-2, 0)$

$$\Rightarrow m_{PQ} = \frac{y_2 - y_1}{x_2 - x_1} = \frac{0 - 0}{3 + 2} = \frac{0}{5} = 0$$

These two points lie on the *x*-axis. Hence the *x*-axis and any line parallel to the *x*-axis has gradient 0.

(iii) $R(0, -4)$, $S(0, 9) \Rightarrow m_{RS} = \frac{y_2 - y_1}{x_2 - x_1} = \frac{9 + 4}{0 - 0} = \frac{13}{0}$

which has no meaning.

These two points lie on the *y*-axis. Hence the gradient of the *y*-axis and any line parallel to the *y*-axis is *undefined*.

Example 2

Show that the points $A(-1, 5)$, $B(1, 0)$ and $C(3, -5)$ are collinear (i.e. lie in the same straight line).

$$m_{AB} = \frac{y_2 - y_1}{x_2 - x_1} = \frac{5 - 0}{-1 - 1} = \frac{5}{-2} = -\frac{5}{2}$$

and $m_{BC} = \frac{0 + 5}{1 - 3} = \frac{5}{-2} = -\frac{5}{2}$

AB and BC have the same gradient and since they have a point in common the 3 points must lie in the same straight line (i.e. be collinear).

ASSIGNMENT 2.1

1. Find, using tables where necessary, the angle which the join of the following pairs of points makes with the positive direction of the *x*-axis.

   (i) $(0, 0)$, $(\sqrt{3}, 1)$ (ii) $(0, 0)$, $(2, 2)$
   (iii) $(3, 2)$, $(1, 3)$ (iv) $(-1, -3)$, $(-4, 1)$
   (v) $(0, 0)$, $(1, -\sqrt{3})$ (vi) $(-5, -6)$, $(-3, -7)$

2. Use the gradient formula to find the gradient of the lines joining the following pairs of points:

   (i) $(5, 6)$, $(6, 7)$ (ii) $(-5, 2)$, $(3, 4)$
   (iii) $(0, 5)$, $(7, 0)$ (iv) $(4, 2\frac{1}{2})$, $(-2, -1)$
   (v) $(-2, -5)$, $(-6, -7)$ (vi) $(6, -3)$, $(-3, 5)$
   (vii) (a, b), $(-a, -b)$ (viii) (x, y), $(x+h, y+k)$

3. A is the point $(-2, -1)$, B is $(2, 0)$, C is $(3, 3)$ and D is $(-1, 2)$. Show that ABCD is a parallelogram.

4. K, L, M, N are the points $(3, 5)$, $(0, 2)$, $(1, 4)$ and $(-5, -2)$ respectively. Show that KL is parallel to MN. Is KM parallel to LN?

5. Which of the following sets of 3 points are collinear?

   (i) $(-7, 4)$, $(-3, 2)$, $(5, -2)$
   (ii) $(-1, 0)$, $(-5, -3)$, $(7, 6)$
   (iii) $(-5, 4)$, $(0, -1)$, $(3, -4)$

(iv) $(-6, -1), (-3, 1), (5, 6)$

   (v) $(1, 6), (7, -1), (5, 2)$

6. The line joining the points $(4, 2)$ and $(-8, 5)$ is parallel to the line joining the points $(5, -6), (a, -2)$. Find the value of a.

7. $P(-2, 5)$, $Q(4, -1)$ and $R(-3, 1)$ are the vertices of a triangle. Find the coordinates of the mid-point of PQ and PR and show that the join of the mid-points of PQ and PR is parallel to QR.

8. Which of the following points can be joined to give parallel lines?

$$(-2, 3), (4, -3), (-3, -1), (5, 1)$$

9. P is the point $(-3, 0)$ and Q is $(-5, -4)$. Which, if any of the following points lie on the line PQ?

$$(3, 5), (-1, 4), (0, 6)$$

Then $l = \{O(0, 0)\} \cup \{P(x, y): m_{OP} = m\}$
$= \{O(0, 0)\} \cup \left\{P(x, y): \dfrac{y}{x} = m, x \neq 0\right\}$
$= \{(x, y): y = mx\}$, since $O(0, 0)$ is the member of this set given by $x = 0$.

The equation $y = mx$ is called the equation of the line l.

Note: The point (a, b) lies on $l \Leftrightarrow b = ma$. This means:

1. If (a, b) lies on l then $b = ma$.

and 2. If $b = ma$ then (a, b) lies on l.

THE EQUATION OF A STRAIGHT LINE IN THE FORM $y = mx$

i.e. the equation $y = mx$ of the line l through the origin and with gradient m.

Let $P(x, y)$ be a point on l distinct from $O(0, 0)$.

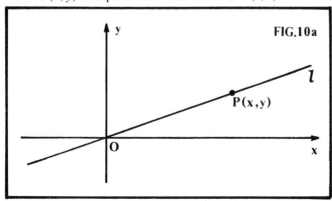

FIG. 10a

THE EQUATION OF A STRAIGHT LINE IN THE FORM $y = mx + c$

2.2

To find the equation of the line l through the point $C(0, c)$ with gradient m, let $P(x, y)$ be a point on l distinct from $C(0, c)$.

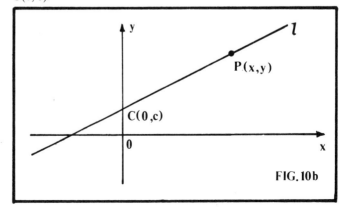

FIG. 10b

Then $l = \{C(0, c)\} \cup \{P(x, y): m_{CP} = m\}$
$= \{C(0, c)\} \cup \left\{P(x, y): \dfrac{y-c}{x} = m, x \neq 0\right\}$
$= \{(x, y): y - c = mx\},$ since $C(0, c)$ is the member of this set given by $x = 0$
$= \{(x, y): y = mx + c\}$

The equation $y = \boldsymbol{mx + c}$ is said to be the equation of the line l.

Note: The point (a, b) lies on $l \Leftrightarrow b = ma + c$. This means:

1. If (a, b) lies on l then $b = ma + c$

and 2. If $b = ma + c$, then (a, b) lies on the line l.

Example 1

Find the equation of the straight line of gradient 2 and passing,

(i) through the point $(0, -3)$

(ii) through the point $(-2, 4)$.

(i) Any straight line of gradient 2 has equation $y = 2x + c$ where c is a constant.
The line passes through
$(0, -3) \Leftrightarrow -3 = 0 + c \Rightarrow c = -3$
and hence the required line has equation $y = 2x - 3$.

(ii) The line passes through
$(-2, 4) \Leftrightarrow 4 = 2 \times (-2) + c$
$\Rightarrow c = 8$
and hence the required line has equation $y = 2x + 8$.

LINES PARALLEL TO THE x OR y-AXIS

In figure 11 let L be a line through the point $H(h, 0)$ and parallel to the y-axis.

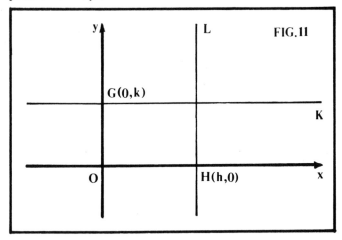

FIG. 11

Let K be a line through the point $G(0, k)$ and parallel to the x-axis.

If $P(x, y)$ be any point on L distinct from $H(h, 0)$.

Then $L = \{(x, y): x = h\}$

Hence any equation of the form $x = h$ represents a line parallel to the y-axis.

Similarly $K = \{(x, y): y = k\}$

Hence any equation of the form $y = k$ represents a line parallel to the x-axis.

THE EQUATION OF THE STRAIGHT LINE IN THE FORM $ax + by + c = 0$

This equation when a and b are not both zero represents a straight line.

(i) If a and b are both not zero, then
$$ax+by+c = 0 \Rightarrow by = -ax-c$$
$$\Rightarrow y = -\frac{a}{b}x - \frac{c}{b}$$

which is of the form $y = mx+c$ and therefore represents a straight line. The gradient of the line $ax+by+c = 0$ is $-\frac{a}{b}$ and $\left(0, -\frac{c}{b}\right)$ is a member of the solution set.

(ii) If $a = 0, b \neq 0$, then $by+c = 0 \Leftrightarrow y = -\frac{c}{b}$,
i.e. a straight line parallel to the x-axis.

(iii) If $b = 0, a \neq 0$, then $ax+c = 0 \Leftrightarrow x = -\frac{c}{a}$,
i.e. a straight line parallel to the y-axis.

(iv) If $c = 0, b \neq 0$, then $ax+by = 0$
$$\Leftrightarrow by = -ax$$
$$\Leftrightarrow y = -\frac{a}{b}x, \text{ i.e. } y = mx$$

i.e. a straight line through the origin and with gradient
$$-\frac{a}{b}$$

Example 2

Find the equation of the line through the point $(2, -3)$ and parallel to the line $2x-3y+4 = 0$.

Note: Since the line $ax+by+c = 0$ can be written
$$y = -\frac{a}{b}x - \frac{c}{b}$$
then this line has gradient $-\frac{a}{b}$.

$\therefore$ Any line parallel to this line may be written
$$y = -\frac{a}{b}x + \frac{k}{b}$$
$$\Rightarrow by = -ax + k$$
$$\Rightarrow ax+by = k$$

Hence any line parallel to the line $2x-3y+4 = 0$ will have equation $2x-3y = k$.
The point $(2, -3)$ lies on the line
$$\Leftrightarrow 4+9 = k$$
$$\Leftrightarrow k = 13$$

Hence the required line has equation
$$2x-3y = 13 \quad \text{or} \quad 2x-3y-13 = 0$$

ASSIGNMENT 2.2

1. Write down the equation of the line,
   (i) having gradient 2 and passing through the origin
   (ii) having gradient $-\frac{1}{3}$ and passing through the origin

2. Write down the equation of the line passing,
   (i) through the origin and the point $(2,2)$
   (ii) through the origin and the point $(-3,4)$
   (iii) through the origin and the point $(0, -\frac{1}{4})$

3. Write down the gradients of the following lines and the coordinates of the points where they cut the y-axis:
   (i) $y = -3x+4$ (ii) $y = \frac{1}{2}x - 1$
   (iii) $2y = 6x - 5$ (iv) $3y = 9x + 8$
   (v) $3y = -x + 6$ (vi) $2y + x = 4$
   (vii) $2x - y = 7$ (viii) $6x + 3y - 4 = 0$

4. Which of the following pairs of lines are parallel?
   (i) $2x-3y=4$
   $x-y=2$
   (ii) $x+4y=7$
   $2x+8y=3$
   (iii) $2y=6x+3$
   $y=3x-1$

5. The straight lines $6x-4y+3=0$ and $3x-ay+8=0$ are parallel. Find a.

6. Find a relationship between a, b, p and q so that the lines $ax+by+c=0$ and $px+qy+r=0$ will be parallel.

7. Write down the equation of the straight line
   (i) having gradient -2 and passing through the point $(0, 2)$
   (ii) having gradient 3 and passing through the point $(2, -1)$
   (iii) having gradient $-\frac{1}{2}$ and passing through the point $(4, 3)$

8. Sketch on the same diagram the set of lines given by
   $\{(x, y): y = 2x+c\}$, where $c \in \{-2, -1, 0, 1, 2\}$

9. Sketch on the same diagram the set of lines given by
   $\{(x, y): y = mx+3\}$, where $m \in \{-1, -\frac{1}{2}, 0, \frac{1}{2}, 1, 2\}$

10. Find the equations of the straight lines through the given points and parallel to the given lines.
    (i) $(2, -1)$; $2x-3y+1=0$
    (ii) $(-2, -4)$; $x+4y+6=0$
    (iii) $(0, -4)$; $5y = -2x+4$
    (iv) $(1, -3)$; $2x+3y=6$
    (v) $(5, -3)$; $3x=4y-6$

11. Give the equation of the straight line through the point $(-2, 3)$ and parallel to,
    (i) the y-axis
    (ii) the x-axis

12. Write down the set of points given by,
    (i) $\{(x, y): x = 4\} \cap \{(x, y): y = -1\}$
    (ii) $\{(x, y): 2y-x+3 = 0\} \cap \{(x, y): x = -3\}$
    Draw a sketch to show each result.

13. Give the equation of the straight line through the origin and parallel to the following lines,
    (i) $2x-y = 4$
    (ii) $4y-2x+3 = 0$
    (iii) $x-3y+6 = 0$
    (iv) $\frac{y}{3} - \frac{x}{4} = 1$

14. Which of the following equations represent straight lines?
    (i) $x+y = 1$
    (ii) $x^2+y = 0$
    (iii) $\frac{x}{2} + \frac{y}{3} = 1$
    (iv) $3x = -4y$
    (v) $x^2+y^2 = 2$
    (vi) $xy = 9$
    (vii) $7 = 3y$
    (viii) $x+y = 0$
    (ix) $\frac{2}{3}y - \frac{1}{4}x = \frac{1}{2}$

15. Find the equation of the straight line through the point $(-2, 3)$ and of gradient 2.

16. Find p, so that the point $(p, -2)$ lies on the line with equation $2x-y = 8$.

17. The point with coordinates $P(p, q)$ lies on the lines with equations $2x-y = 4$ and $x-3y = -3$. Find the co-ordinates of the point P. Does the point P lie on the line $4x+y-14 = 0$? If so what can you say about the three lines?

18. Which, if any, of the following points lie on the line with equation $2x-3y+1 = 0$;
    $(-3, 2), (4, 3), (1, 1), (-1, -1), (-5, -3)$?

THE EQUATION OF A STRAIGHT LINE IN THE FORM $y - b = m(x - a)$ 2.3

To find the equation of the straight line l of gradient m and passing through the point (a, b):

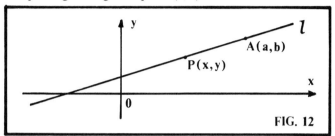

FIG. 12

in figure 12 let $P(x, y)$ be any other point on the line l with gradient m, distinct from $A(a, b)$.

Then $l = \{A(a,b)\} \cup \{P(x,y): m_{AP} = m\}$

$= \{A(a,b)\} \cup \left\{P(x,y): \dfrac{y-b}{x-a} = m, x \neq a\right\}$

$= \{(x, y): y - b = m(x - a)\}$, since $A(a, b)$ is the member of this set given by $x = a$.

i.e. the line l of gradient m and passing through the point (a, b) has equation $y - b = m(x - a)$.

Example 1

Find the equation of the line of gradient $\tfrac{2}{3}$ and passing through the point $(4, -1)$.

The required line has equation $(y - b) = m(x - a)$, where $m = \tfrac{2}{3}$, and (a, b) is $(4, -1)$.

Hence the required line has equation $(y + 1) = \tfrac{2}{3}(x - 4)$

$\Rightarrow 3(y + 1) = 2(x - 4)$
$\Rightarrow 3y + 3 = 2x - 8$
$\Rightarrow 3y - 2x + 11 = 0$

i.e. the required line has equation $3y - 2x + 11 = 0$.

Example 2

Find the equation of the line through the points $P(3, -2)$ and $Q(-5, -4)$.

$$m_{PQ} = \frac{y_2 - y_1}{x_2 - x_1} = \frac{-4 + 2}{-5 - 3} = \frac{-2}{-8} = \frac{1}{4}$$

Now use $m = \tfrac{1}{4}$ and one of the points, say $P(3, -2)$. Hence the line PQ has equation $(y - b) = m(x - a)$

$\Rightarrow (y + 2) = \tfrac{1}{4}(x - 3)$
$\Rightarrow 4y + 8 = x - 3$
$\Rightarrow 4y - x + 11 = 0$

i.e. the line PQ has equation $4y - x + 11 = 0$.

PERPENDICULAR LINES 2.4

Theorem: If two lines AB and CD of gradients m_1 and m_2 are perpendicular then $m_1 m_2 = -1$.

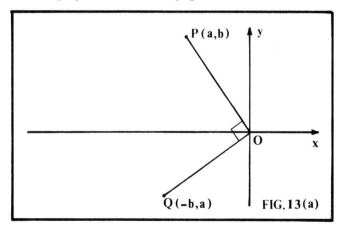

FIG. 13(a)

Proof: Case I (see figure 13(a))
Let the line OP through the points $O(0,0)$ and $P(a,b)$ be rotated anticlockwise through $90°$ about O to a new position OQ.

Under this rotation $P(a,b) \to Q(-b, a)$

and $$m_1 = m_{OP} = \frac{b}{a}$$

$$m_2 = m_{OQ} = \frac{a}{-b}$$

Hence $$m_1 m_2 = \frac{b}{a} \times \frac{a}{-b} = -1$$

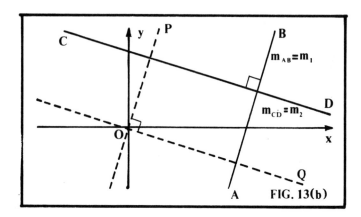

FIG. 13(b)

Case II (see figure 13(b))
If the two lines do not pass through the origin O, then consider the two lines through the origin parallel to the given lines.

$$\text{OP parallel to AB} \Rightarrow m_{OP} = m_{AB} = m_1$$

$$\text{OQ parallel to CD} \Rightarrow m_{OQ} = m_{CD} = m_2$$

But $m_{OP} \cdot m_{OQ} = -1$ (by Case I).

$$\therefore m_1 \cdot m_2 = -1$$

Converse: If $m_1 m_2 = -1$ then AB is perpendicular to CD.

Proof of the Converse:
Here we are given that two lines OP and OQ of gradients m_1 and m_2 are such that $m_1 m_2 = -1$.

Let OR be a line of gradient m_3 perpendicular to OP and in the plane of OPQ.

Then OR perpendicular to OP $\Rightarrow m_1 \cdot m_3 = -1$ (by the theorem).

But $$m_1 \cdot m_2 = -1$$

Hence $$m_2 = m_3$$

i.e. the two lines OQ and OR have the same gradient and must coincide.

Thus OP is perpendicular to OQ.

Hence,

two straight lines of gradient m_1 and m_2 are perpendicular $\Leftrightarrow m_1 \cdot m_2 = -1$

so

(i) If the lines are perpendicular then $m_1 \cdot m_2 = -1$.

(ii) If $m_1 \cdot m_2 = -1$, then the two lines are perpendicular.

Example 1

Find the equation of the line through the point $(-2, 3)$ and perpendicular to the line with equation $3x - 2y + 4 = 0$.

Note the two methods:

(i) line $3x - 2y + 4 = 0 \Leftrightarrow 3x + 4 = 2y$
$$\Leftrightarrow y = \tfrac{3}{2}x + 2$$

and therefore has gradient $\tfrac{3}{2}$. Hence the required line has gradient $-\tfrac{2}{3}$ and has to pass through the point $(-2, 3)$.

∴ the required line has equation $(y - 3) = -\tfrac{2}{3}(x + 2)$
$$\Rightarrow 3(y - 3) = -2(x + 2)$$
$$\Rightarrow 3y - 9 = -2x - 4$$
$$\Rightarrow 2x + 3y - 5 = 0$$

(ii) Since the line $3x - 2y + 4 = 0$ has gradient $\tfrac{3}{2}$, the required line has gradient $= -\tfrac{2}{3}$.

∴ the required line has equation $y = -\tfrac{2}{3}x + c$ and passes through the point $(-2, 3)$
$$\Leftrightarrow 3 = -\tfrac{2}{3} \cdot (-2) + c$$
$$\Leftrightarrow 9 = 4 + 3c$$
$$\Leftrightarrow c = \tfrac{5}{3}$$

hence the required line has equation $y = -\tfrac{2}{3}x + \tfrac{5}{3}$,

i.e. $\qquad 3y = -2x + 5$

i.e. $\qquad 2x + 3y - 5 = 0$

Example 2

The straight line through the point $T(1, 9)$ perpendicular to the line PQ with equation $4x + y + 4 = 0$ meets PQ at R. Find the coordinates of R.

In figure 14, PQ has equation

$$4x + y + 4 = 0 \Leftrightarrow y = -4x - 4$$

Hence $\qquad m_{PQ} = -4 \Rightarrow m_{RT} = \tfrac{1}{4}$

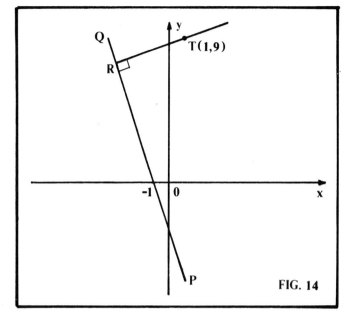

FIG. 14

RT has gradient $\tfrac{1}{4}$ and passes through the point $T(1, 9)$.
Hence RT has equation $y - 9 = \tfrac{1}{4}(x - 1)$
$$\Rightarrow 4y - 36 = x - 1$$
$$\Rightarrow 4y - x = 35$$

Since R lies on both lines

at R $\quad 4y - x = 35$ ⎱ To find the coordinates of R we solve
and $\quad y + 4x = -4$ ⎰ these two equations simultaneously.

$$\Rightarrow \begin{array}{r} 16y - 4x = 140 \\ y + 4x = -4 \\ \hline \end{array}$$
$$\Rightarrow 17y = 136 \Rightarrow y = 8$$

When $y = 8$, $4y - x = 35$
$$\Rightarrow 32 - x = 35 \Rightarrow x = -3$$

Hence the point R has coordinates $(-3, 8)$.

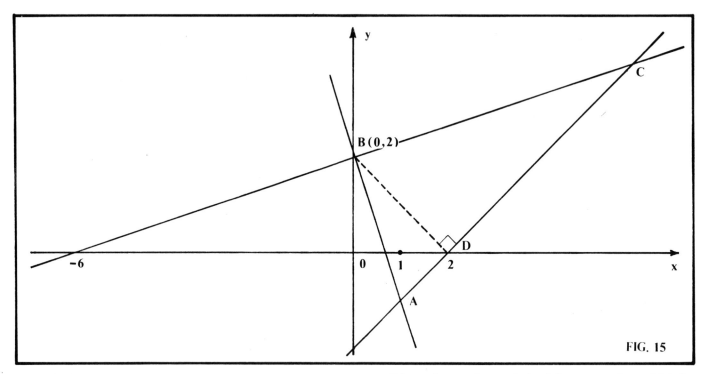

FIG. 15

Example 3

In triangle ABC, the sides AB, BC and CA have equations $3x+y-2=0$, $x-3y+6=0$ and $x-y-2=0$ respectively. Show that angle ABC is a right angle. If BD is an altitude of triangle ABC, prove that D lies on the x-axis and find the length of BD.

In figure 15,
$$AB: 3x+y = 2$$
$$BC: x-3y = -6$$
$$CA: x-y = 2$$

These lines may be drawn quickly by putting $x=0$ and $y=0$ in each equation and obtaining two points which lie on each line and on the axes.

(i) *To show angle ABC is right.*

Line AB has equation $y = -3x+2 \Rightarrow m_{AB} = -3$.
Line BC has equation $3y = x+6 \Rightarrow m_{BC} = \frac{1}{3}$.
Hence $m_{AB} \cdot m_{BC} = -3 \times \frac{1}{3} = -1$
$\Rightarrow$ AB is perpendicular to BC
$\Rightarrow$ angle ABC is a right angle.

Note: Proving a triangle is right angled is much quicker this way, by gradients, than by using the distance formula.

(ii) *BD is an altitude. To prove D lies on the x-axis.*
Line AC has equation $x - y - 2 = 0$.
$$\Rightarrow y = x - 2$$
$$\Rightarrow m_{AC} = 1$$
$$\Rightarrow m_{BD} = -1$$
(since BD is an altitude, BD is perpendicular to AC and therefore $m_{BD} \cdot m_{AC} = -1$).
Also B is the point $(0, 2)$.
Hence BD has equation $y - 2 = -1(x - 0)$.
$$\Rightarrow y - 2 = -x$$
$$\Rightarrow x + y = 2$$
At D, $x + y = 2$ (equation of BD)
and $\quad\;\; x - y = 2$ (equation of AC)
$\quad\;\;\;\;\;\; \overline{2x \;\;\;\;\; = 4}$
$\quad \Rightarrow \quad x = 2$ and hence $y = 0$.
Thus D is the point $(2, 0)$ and lies on the x-axis.

(iii) *Length of BD.*
D is $(2, 0)$, B is $(0, 2)$ and $BD^2 = 2^2 + 2^2 = 8$.
$$\Rightarrow BD = 2\sqrt{2}$$

ASSIGNMENT 2.3

1. Find the equations of the straight lines through the given points and having the given gradients,
   (i) $(4, 3); 3$
   (ii) $(-3, 1); -2$
   (iii) $(-2, -3); -4$
   (iv) $(5, -1); \frac{1}{2}$
   (v) $(-7, 6); -\frac{2}{3}$
   (vi) $(p, q); k$
   (vii) $(\frac{1}{3}, -\frac{2}{3}); -3$

2. Find the equations of the straight lines through the following pairs of points,
   (i) $(0, 2), (6, -3)$
   (ii) $(2, 5), (4, -3)$
   (iii) $(-1, -6), (5, -2)$
   (iv) $(9, -9), (-3, -6)$
   (v) $(-1, 4), (-3, 6)$
   (vi) $(-4, -5), (-8, -3)$

3. Find the equation of the line,
   (i) through the point $(2, -5)$ and parallel to the line $x - 3y + 4 = 0$.
   (ii) through the point $(-3, -4)$ and parallel to the line $2x + y = 9$.

4. P is the point $(-3, 4)$ and $Q(1, -4)$. Find the equation of PQ and the coordinates of the points in which PQ cuts the coordinate axes.

5. State the gradients of the lines perpendicular to the lines with gradients,
   (i) 3
   (ii) -2
   (iii) $\frac{2}{3}$
   (iv) $-\frac{1}{4}$
   (v) 1
   (vi) $-\frac{3}{5}$
   (vii) $1\frac{1}{2}$

6. Find the gradient of the lines perpendicular to the lines with equations,
   (i) $y = 2x + 3$
   (ii) $3y = 2x - 4$
   (iii) $2x + y + 1 = 0$
   (iv) $x + 2y - 1 = 0$
   (v) $6x - y + 2 = 0$
   (vi) $x - 4y + 7 = 0$

7. Find the equations of the lines through the given points and perpendicular to the given lines,
   (i) $(0, 0); 2x - y = 4$
   (ii) $(2, 3); x + y + 2 = 0$
   (iii) $(1, -2); x = 4$
   (iv) $(3, -2); y = -2$

8. Find the equation of the line through the point $(5, -3)$ and perpendicular to the join of $(2, -3)$ and $(-4, -5)$.

9. Find the equation of the perpendicular bisector of the line joining the points P(4, −6) and Q(−2, 6).

10. Show that K(5, 3), L(3, 7), M(−5, 3) and N(−3, −1) are the vertices of a rectangle.

11. Show that the triangle whose vertices are P(0, −1), Q(9, 2) and R(3, −4) is right-angled. Find the equation of the hypotenuse.

12. The straight line joining the points with coordinates (−5, 3) and (−1, −5) is perpendicular to the straight line joining the points (−5, −2) and (a, 1). Find the value of a.

13. Find the equations of the altitudes of the triangle with vertices P(−3, 0), Q(3, 3) and R(2, −2).

14. P(4, −5), Q(3, 0) and R(−2, 1) are three vertices of a rhombus. Find the equations of the diagonals of the rhombus and calculate the coordinates of the fourth vertex.

15. Find, where possible, the coordinates of the point of intersection of the following pair of lines:

    (i) $2x + y - 3 = 0$
    $x + 2y - 3 = 0$

    (ii) $x - 3y + 9 = 0$
    $2x + y - 10 = 0$

    (iii) $2x - 4y - 3 = 0$
    $x - 2y + 9 = 0$

    (iv) $5x - 2y + 7 = 0$
    $3x + 4y - 1 = 0$

    (v) $y = 2x + 4$
    $3y = x - 3$

    Explain why part (iii) has no point of intersection.

16. Show that the following sets of lines are concurrent. (*Note:* Find the point of intersection of two of the lines and show that this point lies on the third line.)

    (i) $x + 2y + 7 = 0$
    $2x - 3y = 0$
    $3x - 2y + 5 = 0$

    (ii) $4x + 3y = 6$
    $x - 5y = 13$
    $3x + y = 7$

17. The sides of a triangle have equations $x + y = 0$, $2x - y + 6 = 0$ and $x + 4y + 12 = 0$. Find the coordinates of the vertices of the triangle.

18. The line through the point A(−5, −7) perpendicular to the line PQ with equation $x + 3y = 4$ meets PQ at B. Find the coordinates of B.

19. The straight line through the point P(0, −3) parallel to the line with equation $x - 3y + 6 = 0$ meets the line through the point Q(2, 1) perpendicular to the line $x - 3y + 6 = 0$ in the point T. Find the coordinates of T.

20. A(−4, 4), B(−3, 1) and C(4, 0) are three vertices of a kite ABCD.

    (i) Calculate the lengths of the sides of the kite.
    (ii) Find the equations of the diagonals.
    (iii) Calculate the point of intersection of the diagonals and hence the coordinates of the vertex D.

21. Show that the points A(0, −2), B(8, 2) and C(6, 6) are three vertices of a rectangle and find the coordinates of the fourth vertex D.

22. P(1, 4), Q(−2, −3) and R(6, −1) are the vertices of a triangle. The median PS meets the altitude QT in K. Find the coordinates of K.

23. Show that the points A(−3, 1), B(−1, −3), C(3, −1) and D(1, 3) are the vertices of a square. The diagonal BD and the line joining A to the mid-point of BC intersect at K. Find the coordinates of K.

ASSIGNMENT 2.4

SUPPLEMENTARY EXAMPLES

1. P, Q and R are the points $(-3, 5)$, $(11, 3)$ and $(-4, -12)$ respectively. The straight line through M, the mid-point of PQ, and perpendicular to QR, meets the straight line through R parallel to PQ at the point S. Find the coordinates of S and verify that the triangle RMS is isosceles.

2. K, L and M are the points $(2, 3)$, $(0, -7)$ and $(8, -3)$ respectively. The perpendicular from K to LM and the perpendicular bisector of KM meet at N. Find the coordinates of N and show that LM = 2MN.

3. A is the point $(0, -8)$ and B the point $(0, -2)$. A point $P(x, y)$ belongs to the set $\{P(x, y): PA = 2PB\}$. Find the equation of the locus of P.

4. P is the point $(-5, 2)$, Q is $(-3, -2)$ and R is $(4, -1)$. The line through R parallel to PQ meets the line through P perpendicular to QR at S; find the co-ordinates of S.

5. $P(-1, -1)$, $Q(2, 1)$ and $S(-6, 2)$ are three points of a parallelogram PQRS. Find the coordinates of R. PS is rotated through an angle of 90° anticlockwise about P to the position PT. Find the coordinates of T.

6. A is the point $(0, 6)$, B is $(-2, -4)$ and C is $(6, 4)$. Find the equations of the altitudes AD, BE and CF and show that the three altitudes are concurrent.

7. A is $(3, 5)$, $B(-3, 1)$ and $C(0, 7)$. Find the equation of the line through the mid-point M of AB parallel to BC. Verify that this line passes through the mid-point N of AC. Show that the median AD bisects MN.

8. P, Q, R and S are the points $(8, 3)$, $(6, 1)$, $(-3, -8)$ and $(0, 7)$ respectively.
   (i) Show that P, Q and R are collinear and calculate the ratio PQ:QR.
   (ii) RQ is produced to T so that RQ = 2QT. Find the coordinates of T.
   (iii) Prove that the straight line through Q perpendicular to PT passes through S.

9. A triangle has vertices $A(7, 1)$, $B(-2, -3)$ and $C(4, 2)$. If D and E are the mid-points of BC and AC respectively, find the equations of the medians AD and BE. Verify that their point of intersection G lies on the third median CF. Find also the ratio in which G divides CF.

10. Show that the triangle ABC with vertices $A(-3, 1)$, $B(1, 2)$ and $C(-4, 5)$ is right-angled. Find the equation of the altitude AD and show that it bisects BC.
    If the altitude AD cuts the y-axis at D, find the coordinates of D and show that CD is parallel to AB. Calculate the area of the triangle ABD.

11. The line $3x - 5y + 15 = 0$ is reflected in the y-axis. Find the equation of the image line. The image line cuts the x-axis at A and the y-axis at B and the line through A perpendicular to the image line cuts the y-axis at C.
    (i) Find the coordinates of C.
    (ii) Calculate the area of triangle ABC.

12. P, Q and R are the points $(5, 3)$, $(-8, 7)$ and $(1, -5)$ respectively. Find the equation of the altitude of the triangle PQR from P to QR. The line through R parallel to the y-axis meets this altitude at K and the line through P parallel to the x-axis meets QR at T. Show that TK is the perpendicular bisector of PR.

13. The straight line with equation $2x - 5y + 17 = 0$ bisects at right angles the line PQ. If P is the point $(1, -2)$

find the coordinates of Q.

14. In triangle ABC, the sides AB, BC and CA have equations
$$3y = x+3, \quad x+y = 1 \quad \text{and} \quad 3x-y = 9$$
respectively.
   (i) Calculate the coordinates of the 3 vertices of the triangle.
   (ii) Prove that the triangle ABC is isosceles.
   (iii) Prove that the perpendicular line from A to BC intersects BC at the point $(\frac{5}{4}, -\frac{1}{4})$.

15. OPQR is a square, the points O, P and Q being respectively $(0,0), (3,0)$ and $(3,3)$. M is the mid-point of QR, N the mid-point of PQ. Show that OM and RN are perpendicular.
   If OM and RN intersect at T show that PT is equal to the length of the square OPQR.

16. P is the point $(3,2)$ and Q is the reflection of P in the line MN with equation $y = 1$. R is the reflection of Q in the line ML with equation $2y+4x=7$.
   (i) Find the coordinates of Q and R.
   (ii) Calculate the length of PR.
   (iii) Show that the line through M perpendicular to PR bisects PR.

ASSIGNMENT 2.5

Objective type items testing sections 2.1–2.4.
Instructions for answering these items are given on page 16.

1. The equation of the line passing through the point $(0, -3)$ and parallel to the line $2x+3y+5 = 0$ is
   A. $3x+2y+6 = 0$
   B. $2x+3y+9 = 0$
   C. $2x+3y+3 = 0$
   D. $y = 2x+3$
   E. $2y-3x+6 = 0$

2. The equation of a line perpendicular to the line with equation $2x-4y+1=0$ is
   A. $3x+6y=1$
   B. $4y=8x+1$
   C. $2y+8x=1$
   D. $4y=2x+1$
   E. $3y+6x=1$

3. The lines $x-2y=4$ and $6x+ay=8$ are perpendicular. The value of a is
   A. -3
   B. -12
   C. 0
   D. 12
   E. 3

4. Two lines are defined by $L_1 = \{(x,y): 2x+y=5\}$ and $L_2 = \{(x,y): 3x-4y=13\}$. $L_1 \cap L_2$ is the set
   A. $\{(3,-1)\}$
   B. $\{(-3,1)\}$
   C. $\emptyset$
   D. $\{(1,3), (4,-\frac{1}{2})\}$
   E. none of these.

5. PQRS is a rectangle with P the point $(-2,6)$ and R the point $(1,-3)$. One of the other vertices of the rectangle could have coordinates
   A. $(0,4)$
   B. $(-1,-4)$
   C. $(3\frac{1}{2}, 4\frac{1}{2})$
   D. $(3,5)$
   E. $(-5,3)$

6. The line $\frac{x}{7} - \frac{y}{2} = 1$ cuts the x-axis at X and the y-axis at Y. The coordinates of X and Y are
   A. $(2,0), (7,0)$
   B. $(-2,0), (0,7)$
   C. $(7,0), (0,-2)$
   D. $(\frac{1}{7}, 0), (0, -\frac{1}{2})$
   E. $(-\frac{1}{2}, 0), (0, \frac{1}{7})$

7. The line $y = mx$ passes through the point of intersection of the lines $x = p$ and $y = q$, if and only if
   A. $m = \frac{p}{q}$
   B. $m = -\frac{p}{q}$
   C. $m = \frac{q}{p}$
   D. $m = -\frac{q}{p}$
   E. m has some other value involving p and q

8. The line l with equation $y = mx + c$ maps onto the line l' on reflection in the x-axis. Which of the following statements is false?
   A. l and l' are equally but oppositely inclined to the x-axis.
   B. l' has equation $y = mx - c$.
   C. Both l and l' cut the x-axis at the point $\left(-\frac{c}{m}, 0\right)$.
   D. If $\tan \theta = m$, then the angle between l and l' is 2θ.
   E. Point $P(a,b)$ on $l \to P'(a,-b)$ on l'.

9. If m is a variable on the set $\{0, 1, 2, 3\}$, then the set $\{(x, y): mx + y = 2\}$ represents,

    (1) a set of 4 parallel straight lines.

    (2) a set of 4 lines passing through the point $(0, 2)$.

    (3) the line $y = 2$ and three other lines perpendicular to the line $y = 2$.

10. If c is a variable on the set $\{0, 1, 2\, 3\}$, then the set $\{(x, y): 2x - y = c\}$ represents,

    (1) a set of 4 parallel straight lines of gradient 2.

    (2) the line $y = 2x$ and three other lines intersecting the line $y = 2x$.

    (3) four parallel lines, one through the point $(0, 0)$, a second through $(0, 1)$, a third through $(0, 2)$ and a fourth through $(0, 3)$.

11. (1) Two lines of gradient p and q are perpendicular.

    (2) $p \neq q$.

12. (1) Line PQ has gradient

$$m_{PQ} = \frac{y_2 - y_1}{0}, y_2 \neq y_1$$

    (2) Line PQ is parallel to the y-axis.

UNIT 3: THE CIRCLE

THE CIRCLE $x^2 + y^2 = r^2$ 3.1

The circle C, with centre the origin and with radius r has equation,

$$x^2 + y^2 = r^2$$

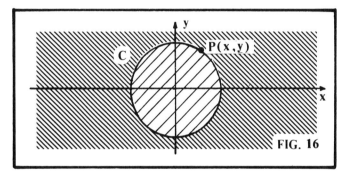

FIG. 16

Let $P(x, y)$ be a point on the circumference of circle C. Then

$$C = \{P : OP = r\}$$
$$= \{P : OP^2 = r^2\}$$
$$= \{(x, y) : x^2 + y^2 = r^2\} \quad \text{(using the distance formula)}$$

Thus by the distance formula the point $P(x, y)$ is always a fixed distance r from a fixed point $(0, 0)$ and therefore lies on a circle centre $(0, 0)$ and radius r.

The equation $x^2 + y^2 = r^2$ is said to be the equation of the circle with centre the origin and radius r.

Note:

(i) $P(a, b)$ lies on the circle $\Leftrightarrow a^2 + b^2 = r^2$

(ii) $P(a,b)$ lies outside the circle $\Leftrightarrow a^2+b^2 > r^2$
(iii) $P(a,b)$ lies inside the circle $\Leftrightarrow a^2+b^2 < r^2$

Example 1

Find the equation of the circle with centre the origin and passing through the point $(2,-4)$.
   Does the point $P(1,5)$ lie inside or outside this circle?
   Any circle with centre the origin and radius r has equation
$$x^2+y^2 = r^2$$
This circle passes through the point $(2,-4)$
$$\Leftrightarrow 2^2+(-4)^2 = r^2$$
$$\Leftrightarrow r^2 = 20$$
Hence the required circle has equation $x^2+y^2 = 20$.

$P(x,y)$ lies outside the circle $\Leftrightarrow x^2+y^2 > 20$
If $\quad P(1,5)$ lies outside the circle $\Leftrightarrow 1^2+5^2 > 20$
and $\quad\quad\quad 1^2+5^2 = 26 > 20$
hence the point $P(1,5)$ lies outside the circle.

Example 2

A is the point $(1,0)$ and B is $(4,0)$. $P(x,y)$ is a member of the set of points whose distance from B is twice its distance from A. Find the equation of the locus of P and interpret your result geometrically.

The distance of $P(x,y)$ from B is twice its distance from A
$$\Leftrightarrow \quad PB = 2PA$$
$$\Leftrightarrow \quad PB^2 = 4PA^2$$
$$\Leftrightarrow (x-4)^2+(y-0)^2 = 4[(x-1)^2+(y-0)^2]$$
(using the distance formula)

$\Leftrightarrow x^2-8x+16+y^2 = 4[x^2-2x+1+y^2]$
$\Leftrightarrow x^2-8x+16+y^2 = 4x^2-8x+4+4y^2$
$\Leftrightarrow \quad 3x^2+3y^2-12 = 0$
$\Leftrightarrow \quad\quad x^2+y^2 = 4$

Hence the locus of P is a circle with centre the origin and radius 2 units.

This example may be written out in set notation thus: Since $P(x,y)$ belongs to the set of points which satisfy the condition $PB = 2PA$, then,

$\{P: PB = 2PA\}$
$= \{P: PB^2 = 4PA^2\}$
$= \{(x,y): (x-4)^2+(y-0)^2 = 4[(x-1)^2+(y-0)^2]\}$
$= \{(x,y): x^2-8x+16+y^2 = 4(x^2-2x+1+y^2)\}$
$= \{(x,y): x^2+y^2-8x+16 = 4x^2+4y^2-8x+4\}$
$= \{(x,y): 3x^2+3y^2-12 = 0\}$
$= \{(x,y): x^2+y^2 = 4\}$

Hence the locus of P is a circle with centre the origin and radius 2 units.

Example 3

Find the points of intersection of the line $x+y = 3$ and the circle $x^2+y^2 = 5$.
$$x+y = 3 \Leftrightarrow y = 3-x$$
The line $y = 3-x$ cuts the circle $x^2+y^2 = 5$
$\Leftrightarrow \quad x^2+(3-x)^2 = 5$
$\Leftrightarrow x^2+9-6x+x^2 = 5$
$\Leftrightarrow \quad 2x^2-6x+4 = 0$
$\Leftrightarrow \quad x^2-3x+2 = 0$
$\Leftrightarrow \quad (x-2)(x-1) = 0$
$\Leftrightarrow \quad\quad x = 2 \quad\text{or}\quad x = 1$

when $x = 2$, $y = 1$ and when $x = 1$, $y = 2$, i.e. the line cuts the circle at the points $(2, 1)$ and $(1, 2)$.

ASSIGNMENT 3.1

1. Find the equations of the circles with centre the origin O and
   (i) of radius 2
   (ii) of radius $2\sqrt{3}$
   (iii) of radius 1·5
   (iv) of radius k
   (v) passing through the point $(4, 2)$
   (vi) passing through the point $(3, -3)$
   (vii) passing through the point $(-5, -4)$
   (viii) passing through the point (a, b)
   (ix) passing through the point $(a\cos\theta, a\sin\theta)$

2. The point $(k, -3)$ lies on the circle $x^2 + y^2 = 25$. Find the coordinates of the point.

3. Which of the following points lie on, lie outside or lie inside the circle with equation $x^2 + y^2 = 29$?

   $(-2, 5)$, $(3, -4)$, $(5, 2)$, $(1, -3)$, $(0, -9)$,
   $(4, -4)$, $(-6, 1)$, $(-5, -2)$

4. The circle with equation $x^2 + y^2 = 36$ divides the x, y plane into three regions A, B and C defined by,

   $A = \{(x, y): x^2 + y^2 < 36\}$
   $B = \{(x, y): x^2 + y^2 = 36\}$
   $C = \{(x, y): x^2 + y^2 > 36\}$

   In which region does each of the following points lie?
   (i) $(-4, 2)$ (ii) $(5, -4)$ (iii) $(0, 6)$
   (iv) $(0, 0)$ (v) $(3\sqrt{2}, 3\sqrt{2})$ (vi) $(-6, 0)$
   (vii) $(-4, -4)$ (viii) $(-5, -5)$ (ix) $(3, 5)$

5. Find the coordinates of the points in which the following circles cut the x and y axes:
   (i) $x^2 + y^2 = 1$
   (ii) $x^2 + y^2 = 9$
   (iii) $x^2 + y^2 = 12$
   (iv) $x^2 + y^2 = 72$

6. Find the points of intersection of the given circle with the given line:
   (i) $x^2 + y^2 = 18$, $y = x$
   (ii) $x^2 + y^2 = 32$, $x + y = 0$
   (iii) $x^2 + y^2 = 10$, $x + y = 2$
   (iv) $x^2 + y^2 = 29$, $x - y + 3 = 0$

7. A is the point $(0, -1)$ and $B(0, -4)$. $P(x, y)$ belongs to the set of points such that $PB = 2PA$. Find the equation of the locus of P and interpret the result geometrically.

8. Find the equation of the locus and interpret the result geometrically of the set of points given by $\{P(x, y): PA = 3PB\}$ where A is $(9, 0)$ and B is $(1, 0)$.

9. The point $(-1, 7)$ lies on the circle $x^2 + y^2 = 50$. Write down three other points (h, k) that lie on this circle, $h, k \in Z$.

10. Find the equation of the circle touching the sides of the triangle with sides $y + x = 2$, $y - x = 2$ and $y = -\sqrt{2}$.

11. A circle has centre the origin and touches the line with equation $x + y = 4$. Find the equation of the circle.

12. If the radius of the circle with equation $x^2 + y^2 = 9$ is doubled, what would be the equation of the new circle formed?

13. If the radius of the circle with equation $x^2 + y^2 = 20$ is halved, what would be the equation of the new circle formed?

3.2 THE CIRCLE $(x-a)^2 + (y-b)^2 = r^2$

Let $P(x, y)$ be any point on the circumference of circle C, with centre $A(a, b)$ and radius r.

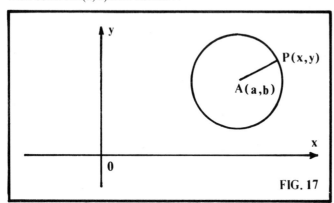

FIG. 17

Then
$$C = \{P : AP = r\}$$
$$= \{P : AP^2 = r^2\}$$
$$= \{(x, y) : (x-a)^2 + (y-b)^2 = r^2\}$$
(using the distance formula)

thus by the distance formula $P(x, y)$ is always a fixed distance r from a fixed point (a, b) and therefore lies on a circle centre (a, b) and radius r.

The equation $(x-a)^2 + (y-b)^2 = r^2$ is said to be the equation of the circle with centre (a, b) and radius r.

Example 1

Find the equation of the circle centre $(-2, 3)$ and passing through the point $(1, -2)$.

Does the point $(4, 1)$ lie inside or outside the circle?

Any circle with centre $(-2, 3)$ and radius r has equation
$$(x+2)^2 + (y-3)^2 = r^2$$

This circle passes through the point $(1, -2)$
$$\Leftrightarrow (1+2)^2 + (-2-3)^2 = r^2$$
$$\Leftrightarrow 9 + 25 = r^2$$
$$\Leftrightarrow r^2 = 34$$

Hence the circle has equation $(x+2)^2 + (y-3)^2 = 34$. Assuming the point $(4, 1)$ lies outside the circle
$$\Leftrightarrow (4+2)^2 + (1-3)^2 > 34$$
$$\Leftrightarrow 36 + 4 > 34$$
$$\Leftrightarrow 40 > 34$$

hence the point $(4, 1)$ lies outside the circle.

Example 2

Find the equation of the circle with centre $(4, -3)$ and radius 5. Find also the points in which this circle cuts the coordinate axes.

Any circle centre (a, b) and radius r has equation
$$(x-a)^2 + (y-b)^2 = r^2$$

Hence the circle centre $(4, -3)$, radius 5 has equation
$$(x-4)^2 + (y+3)^2 = 25$$

Note: It is usual to square out the brackets and express the equation of the circle without the brackets.

Hence the circle has equation
$$x^2 - 8x + 16 + y^2 + 6y + 9 = 25$$
i.e.
$$x^2 + y^2 - 8x + 6y = 0$$

This circle cuts the x-axis at the points where $y = 0$, i.e. at the points where $x^2 - 8x = 0$.
i.e.
$$x(x-8) = 0 \Rightarrow x = 0 \text{ or } 8$$
i.e. at the points $(0, 0)$ and $(8, 0)$.

The circle cuts the y-axis at the points where $x = 0$, i.e. at the points where $y^2 + 6y = 0$.

i.e. $\quad y(y+6) = 0 \Rightarrow y = 0 \text{ or } -6$

i.e. the points $(0,0)$ and $(0,-6)$.

Hence the circle cuts the axes at the points $(0,0)$, $(8,0)$ and $(0,-6)$.

ASSIGNMENT 3.2

1. Write down, in the form $(x-a)^2 + (y-b)^2 = r^2$, the equations of the circles with given centres and radii.
   (i) centre $(1,2)$, radius 3.
   (ii) centre $(-1,-2)$, radius 3.
   (iii) centre $(0,4)$, radius 5.
   (iv) centre $(-5,2)$, radius 6.
   (v) centre $(-3,-3)$, radius 4.
   (vi) centre $(\frac{1}{2}, -\frac{1}{2})$, radius 2.

2. Write down the coordinates of the centre and the length of the radius of each of the following circles:
   (i) $(x-3)^2 + (y-2)^2 = 4$
   (ii) $(x+3)^2 + (y+1)^2 = 9$
   (iii) $(x-2)^2 + (y+5)^2 = 7$
   (iv) $(x+5)^2 + (y-4)^2 = 24$

3. A circle has equation $(x-3)^2 + (y+4)^2 = 25$. Which of the following points lie on the circle, lie inside the circle or lie outside the circle?
   (i) $(-2,-1)$ (ii) $(3,0)$ (iii) $(-4,-1)$
   (iv) $(0,0)$ (v) $(7,-1)$ (vi) $(2,-1)$
   (vii) $(-1,-1)$

4. The circle with centre $(4,-2)$ passes through the point $(-3,-2)$. Find the length of its radius.

5. The circle with centre $(-5,-1)$ passes through the point $(1,-4)$. Find the length of its radius.

6. A circle with centre $(2,7)$ touches the x-axis. Find its equation.

7. A circle touches the y-axis at the point $(0,3)$ and the x-axis at the point $(-3,0)$. Find its equation.

8. In this question, give the equation of the circle in the form $(x-a)^2 + (y-b)^2 = r^2$ and then in the form not containing brackets [for example the circle with equation $(x-2)^2 + (y-1)^2 = 4$, may also be written as $x^2 - 4x + 4 + y^2 - 2y + 1 = 4$, i.e.
$$x^2 + y^2 - 4x - 2y + 1 = 0$$
when the brackets are removed].

   Find the equation (two forms for each) of the circles with,
   (i) centre $(5,-2)$ and radius 3.
   (ii) centre $(-2,-1)$ and radius 9.
   (iii) centre $(3,-4)$ and passing through the origin.
   (iv) centre $(-3,-2)$ and passing through the point $(2,2)$.
   (v) centre $(2,2)$ and passing through the point $(-3,-2)$.

9. Circles are drawn with the following pairs of points as diameter. Find the equations of the circles.
   (i) $P(5,3)$ and $Q(-1,3)$
   (ii) $A(-2,-3)$ and $B(-8,-7)$
   (iii) $O(0,0)$ and $T(6,-4)$

10. Find the equation of the diameter of the circle $(x-3)^2 + (y+1)^2 = 20$ through the point $(1,3)$. Find also the coordinates of the other extremity of this diameter.

11. The circle with centre $(-1, 5)$ touches the y-axis. Find its equation. Find also the equation of the image of this circle in

 (i) the y-axis, (ii) the x-axis, (iii) the origin.

3.3 THE EQUATION $x^2 + y^2 + 2gx + 2fy + c = 0$

We have said that any circle with equation

$$(x+2)^2 + (y-3)^2 = 16$$

say, may be written as

$$x^2 + 4x + 4 + y^2 - 6y + 9 - 16 = 0$$

i.e. $\qquad x^2 + y^2 + 4x - 6y - 3 = 0$

when the brackets are removed.

We now prove the converse, namely that any equation of the form $x^2 + y^2 + 2gx + 2fy + c = 0$, where g, f and c are constants, represents a circle with centre $(-g, -f)$ and radius $\sqrt{(g^2 + f^2 - c)}$ provided $g^2 + f^2 - c > 0$.

Proof: If $P(x, y)$ is a member of the set of points which satisfies the equation $x^2 + y^2 + 2gx + 2fy + c = 0$, then

$\{(x, y): x^2 + y^2 + 2gx + 2fy + c = 0\}$
$= \{(x, y): x^2 + 2gx + g^2 + y^2 + 2fy + f^2 = g^2 + f^2 - c\}$
$= \{(x, y): (x + g)^2 + (y + f)^2 = g^2 + f^2 - c\}$
$= \{(x, y): [x - (-g)]^2 + [y - (-f)]^2 = g^2 + f^2 - c\}$

If C is the point $(-g, -f)$ and $P(x, y)$ any point which satisfies the given equation then by the distance formula $CP^2 = [x - (-g)]^2 + [y - (-f)]^2$ and hence,

$\{(x, y): x^2 + y^2 + 2gx + 2fy + c = 0\}$
$\qquad = \{P: CP^2 = g^2 + f^2 - c\}$

thus the point $P(x, y)$ is always a fixed distance from the fixed point $(-g, -f)$ and therefore lies on the circumference of a circle centre $(-g, -f)$ and radius $CP = \sqrt{(g^2 + f^2 - c)}$,

provided $g^2 + f^2 - c > 0$.

Thus the equation $x^2 + y^2 + 2gx + 2fy + c = 0$, g, f, c constants, represents a circle with centre $(-g, -f)$ and radius $\sqrt{(g^2 + f^2 - c)} \Leftrightarrow g^2 + f^2 - c > 0$.

Note:

(i) In the equation $x^2 + y^2 + 2gx + 2fy + c = 0$, called the general equation of the circle,

 (a) the coefficients of x^2 and y^2 are equal;

 (b) there is no term in xy.

 Thus $3x^2 + 3y^2 + 6x - 2y + 9 = 0$ represents a circle since the coefficients of x^2 and y^2 are equal and there is no term in xy.

(ii) The point (h, k) lies on the circle
$\Leftrightarrow h^2 + k^2 + 2gh + 2fk + c = 0$
 The point (h, k) lies inside the circle
$\Leftrightarrow h^2 + k^2 + 2gh + 2fk + c < 0$
 The point (h, k) lies outside the circle
$\Leftrightarrow h^2 + k^2 + 2gh + 2fk + c > 0$

(iii) If $c = 0$ the circle passes through the origin;
 If $f = 0$ the circle has its centre on the x-axis;
 If $g = 0$ the circle has its centre on the y-axis.

Example 1

Find the coordinates of the centre and the length of the radius of the circle with equation $x^2 + y^2 + 6x - 2y - 6 = 0$.

The circle $x^2 + y^2 + 2gx + 2fy + c = 0$ has centre $(-g, -f)$ and radius $\sqrt{(g^2 + f^2 - c)}$.

Hence the circle $x^2 + y^2 + 6x - 2y - 6 = 0$ has centre $(-3, 1)$ and radius $\sqrt{(9 + 1 + 6)} = \sqrt{16} = 4$.

Or Write $\qquad x^2 + y^2 + 6x - 2y - 6 = 0$

as $\qquad x^2 + 6x + 9 + y^2 - 2y + 1 = 9 + 1 + 6$

i.e. we add to each side of the equation the square of half the coefficient of x and the square of half the coefficient of y.

The equation then becomes
$$(x+3)^2+(y-1)^2 = 16$$
i.e. of the form
$$(x-a)^2+(y-b)^2 = r^2$$
and hence the centre is $(-3, 1)$ and radius 4.

Example 2

Find the centre and radius of the circle
$$2x^2+2y^2-7x+5y+3 = 0$$

$2x^2+2y^2-7x+5y+3 = 0 \Leftrightarrow x^2+y^2-\tfrac{7}{2}x+\tfrac{5}{2}y+\tfrac{3}{2} = 0$

The circle $x^2+y^2+2gx+2fy+c = 0$ has centre $(-g, -f)$ and radius $\sqrt{(g^2+f^2-c)}$.

Hence the circle $x^2+y^2-\tfrac{7}{2}x+\tfrac{5}{2}y+\tfrac{3}{2} = 0$ has centre $(\tfrac{7}{4}, -\tfrac{5}{4})$ and radius

$$\sqrt{(\tfrac{49}{16}+\tfrac{25}{16}-\tfrac{3}{2})} = \sqrt{(\tfrac{50}{16})} = \frac{5\sqrt{2}}{4}$$

Example 3

Find the equation of the circle through the points $(-2, 0)$, $(-3, -2)$ and $(-4, 1)$.

Method 1: Let the circle have equation
$$x^2+y^2+2gx+2fy+c = 0$$

$(-2, 0)$ lies on the circle $\Leftrightarrow 4+0-4g+c = 0$
$\Leftrightarrow -4g+c = -4$

$(-3, -2)$ lies on the circle $\Leftrightarrow 9+4-6g-4f+c = 0$
$\Leftrightarrow -6g-4f+c = -13$

$(-4, 1)$ lies on the circle $\Leftrightarrow 16+1-8g+2f+c = 0$
$\Leftrightarrow -8g+2f+c = -17$

Hence we have
$$\begin{aligned}4g-c &= 4 &(1)\\ 6g+4f-c &= 13 &(2)\\ 8g-2f-c &= 17 &(3)\end{aligned}$$

subtract (1) from (2): $\quad 2g+4f = 9 \quad (4)$
subtract (1) from (3): $\quad 4g-2f = 13 \quad (5)$

from (4) $\quad 2g+4f = 9$
from (5) $\quad 8g-4f = 26$
$\qquad\qquad\quad 10g = 35 \Rightarrow g = \tfrac{7}{2}$

from (4) $\quad 2g+4f = 9 \Leftrightarrow 4f = 9-2g$
$\Leftrightarrow 4f = 9-7 = 2$
$\Rightarrow f = \tfrac{1}{2}$

from (1) $\quad 4g-c = 4 \Leftrightarrow c = 4g-4$
$\Rightarrow c = 14-4 = 10$

i.e. $\quad g = \tfrac{7}{2}, \quad f = \tfrac{1}{2} \quad \text{and} \quad c = 10$

Hence the circle has equation $x^2+y^2+7x+y+10 = 0$.

Method 2: Let A be the point $(-2, 0)$, B the point $(-3, -2)$ and C the point $(-4, 1)$.

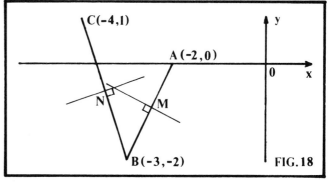

FIG. 18

The centre of the circle lies on the perpendicular bisector of AB and BC.

Now $m_{AB} = \dfrac{0+2}{-2+3} = 2$

The mid-point of AB is $M(-\tfrac{5}{2}, -1)$ and the perpendicular bisector through M has gradient $-\tfrac{1}{2}$.

$\therefore$ the perpendicular bisector of AB has equation

$y+1 = -\tfrac{1}{2}(x+\tfrac{5}{2}) \Rightarrow 4y+4 = -2x-5$
$\Rightarrow 4y+2x = -9 \qquad (1)$

Also $m_{BC} = \dfrac{1+2}{-4+3} = \dfrac{3}{-1} = -3$

and the mid-point of BC is $N(-\tfrac{7}{2}, -\tfrac{1}{2})$ and the perpendicular bisector through N has gradient $\tfrac{1}{3}$.

$\therefore$ the perpendicular bisector of BC has equation

$y+\tfrac{1}{2} = \tfrac{1}{3}(x+\tfrac{7}{2}) \Rightarrow 6y+3 = 2x+7$
$\Rightarrow 6y-2x = 4 \qquad (2)$

Hence at the centre of the circle

$\phantom{\text{and}}\quad 4y+2x = -9 \qquad (1)$
and $\quad 6y-2x = 4 \qquad (2)$
$\Rightarrow 10y = -5 \Rightarrow y = -\tfrac{1}{2}$

from (1) $4y+2x = -9 \Leftrightarrow 2x = -9-4y$
$\Rightarrow 2x = -9+2 = -7$
$\Rightarrow x = -\tfrac{7}{2}$

Hence the centre of the circle is $(-\tfrac{7}{2}, -\tfrac{1}{2})$ and the circle passes through $(-2, 0)$,

$\therefore$ Radius$^2 = (-\tfrac{7}{2}+2)^2 + (-\tfrac{1}{2}-0)^2 = \tfrac{9}{4}+\tfrac{1}{4} = \tfrac{5}{2}$

and the circle has equation

$(x+\tfrac{7}{2})^2 + (y+\tfrac{1}{2})^2 = \tfrac{5}{2} \Rightarrow x^2+7x+\tfrac{49}{4}+y^2+y+\tfrac{1}{4} = \tfrac{5}{2}$
$\Rightarrow x^2+y^2+7x+y+\tfrac{50}{4}-\tfrac{5}{2} = 0$
$\Rightarrow x^2+y^2+7x+y+10 = 0$

ASSIGNMENT 3.3

1. Which of the following equations represent circles?
   (i) $x^2+y^2-6x+6y+3 = 0$
   (ii) $x^2+y^2-2y+1 = 0$
   (iii) $4x^2+4y^2+2y-6x+3 = 0$
   (iv) $2x^2+3y^2-2x+6y+9 = 0$
   (v) $2x^2-2y^2-3x+2xy+4 = 0$
   (vi) $3x^2+3y^2 = 16$
   (vii) $x^2+y^2 = 10x-14y$
   (viii) $x^2+y^2 = 2y$

2. State the coordinates of the centre and find the length of the radius of the following circles:
   (i) $x^2+y^2+2x+8y-8 = 0$
   (ii) $x^2+y^2-6x+6y+2 = 0$
   (iii) $x^2+y^2+4x+10y+13 = 0$
   (iv) $x^2+y^2+6x+8y = 0$
   (v) $x^2+y^2-5x-9y+2 = 0$
   (vi) $2x^2+2y^2+7x-3y-1 = 0$
   (vii) $4x^2+4y^2-12x+16y-11 = 0$
   (viii) $x^2+y^2+2x\cos\theta+2y\sin\theta = 0$
   (ix) $x^2+y^2+4ax+4ay-a^2 = 0$

3. Which of the following points lie on the circle $x^2+y^2-8x+4y-9 = 0$?
   (i) $(3, 0)$ (ii) $(4, -1)$ (iii) $(2, 3)$
   (iv) $(3, 2)$ (v) $(-1, 2)$

4. The point $(-2, k)$ lies on the circle
$$x^2+y^2-6x+3y-20 = 0$$
Find the value of k.

5. Find the coordinates of the points in which the circle with centre $(2, -3)$ and radius 5 cuts the x-axis.

6. Find the equation of the circle whose centre is the point $(-1, 3)$ and which passes through the centre of the circle $x^2 + y^2 - 4x + 2y - 4 = 0$.

7. Show that the circles with equations
$$x^2 + y^2 - 10x - 2y + 10 = 0$$
and $$x^2 + y^2 - 4x + 6y + 12 = 0$$
touch externally.

(*Note*: You have to show that the sum of the radii of the circles is equal to the distance between their centres.)

8. Show that the circles with equations
$$x^2 + y^2 + 2x - 2y - 48 = 0$$
and $$x^2 + y^2 + 6x + 2y - 8 = 0$$
touch internally.

(*Note*: You have to show that the difference of the radii of the circles is equal to the distance between their centres.)

9. A circle passes through the origin and cuts the co-ordinate axes at $(0, 3)$ and $(4, 0)$. Find its equation.

10. Find the equations of the circles through the following points:
    (i) $(0, 0), (0, 3), (4, -1)$
    (ii) $(1, -1), (2, 1), (3, -2)$
    (iii) $(-2, 1), (-3, -1), (-4, 2)$
    (iv) $(-2, -3), (-4, 1), (-2, 1)$

11. Find the equations of the circles with a diameter joining the points,
    (i) $(0, 0)$ and $(-4, 2)$ (ii) $(-2, -1)$ and $(4, 7)$
    (iii) $(2, 5)$ and $(3, 7)$ (iv) $(3, 1)$ and $(-1, 2)$

12. Find the equation of the circle whose centre lies on the y-axis and which passes through the points $(-2, -1)$ and $(2, 0)$.

13. Find the equation of the circle whose centre lies on the x-axis and which passes through the points $(4, 4)$ and $(-1, 4)$.

14. $P(x, y)$ is a member of the set of points whose distance from the origin is twice their distance from the point $(0, -2)$. Find the equation of the locus of P and show that it is a circle whose centre lies on the y-axis.

INTERSECTION OF A LINE AND CIRCLE— TANGENTS   3.4

A line and a circle may,

(i) intersect in two distinct points, as in figure 19(a).

(ii) intersect in one point as in figure 19(b). In this case we say the line intersects or touches the circle in **two coincident points** and we say the line is a **tangent** to the circle. The point of touching or intersection is called the **point of contact**.

(iii) not intersect in any point, as in figure 19(c).

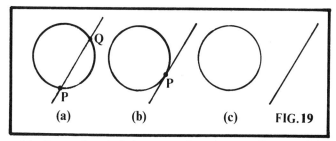

FIG. 19

If a line and a circle intersect then the coordinates of the point or points of intersection will satisfy the equation of the line and the equation of the circle.

Example 1

Find the coordinates of the points of intersection of the line $x-y+2=0$ and the circle $x^2+y^2-4x+2y-12=0$.

$$x-y+2=0 \Leftrightarrow y=x+2$$

The line $y=x+2$ intersects the circle

$$x^2+y^2-4x+2y-12=0$$
$$\Leftrightarrow x^2+(x+2)^2-4x+2(x+2)-12=0$$
$$\Leftrightarrow x^2+x^2+4x+4-4x+2x+4-12=0$$
$$\Leftrightarrow 2x^2+2x-4=0$$
$$\Leftrightarrow x^2+x-2=0$$
$$\Leftrightarrow (x+2)(x-1)=0$$
$$\Rightarrow x=-2 \text{ or } 1$$

When $x=1$, $y=x+2 \Rightarrow y=3$

When $x=-2$, $y=x+2 \Rightarrow y=0$

hence the line intersects the circle at the points with co-ordinates $(1,3)$ and $(-2,0)$.

Example 2

Show that the line $2y-x+8=0$ is a tangent to the circle $x^2+y^2+2x+4y=0$, and write down the coordinates of the point of contact.

$$2y-x+8=0 \Leftrightarrow x=2y+8$$

The line $x=2y+8$ intersects the circle

$$x^2+y^2+2x+4y=0$$
$$\Leftrightarrow (2y+8)^2+y^2+2(2y+8)+4y=0$$
$$\Leftrightarrow 4y^2+32y+64+y^2+4y+16+4y=0$$
$$\Leftrightarrow 5y^2+40y+80=0$$
$$\Leftrightarrow y^2+8y+16=0$$
$$\Leftrightarrow (y+4)(y+4)=0$$
$$\Rightarrow y=-4 \quad \text{(twice)}$$

Since the line intersects the circle at one point (or two coincident points) where $y=-4$, the line is a tangent to the circle.

When $y=-4$, $x=2y+8 \Rightarrow x=-8+8=0$

i.e. the point of contact is $(0,-4)$.

Example 3

Find the equation of the tangent from the point $(0,5)$ to the circle $x^2+y^2=16$.

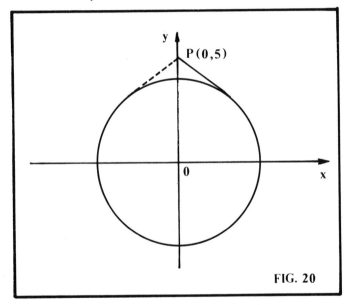

FIG. 20

The line through the point $(0,5)$ with gradient m has equation $y-5=m(x-0)$, i.e. $y=mx+5$.

This line $y=mx+5$ intersects the circle $x^2+y^2=16$,

$$\Leftrightarrow x^2+(mx+5)^2=16$$
$$\Leftrightarrow x^2+m^2x^2+10mx+25=16$$
$$\Leftrightarrow x^2(1+m^2)+10mx+9=0$$

This equation gives the x-coordinates of the points in which the line cuts the circle. If the line is to be a tangent then this equation must have equal roots in x.

The equation $ax^2 + bx + c = 0$ has equal roots in
$$x \Leftrightarrow b^2 - 4ac = 0$$
i.e. its discriminant is zero.

Hence the equation $(1+m^2)x^2 + 10mx + 9 = 0$ has equal roots in x,

$$\Leftrightarrow 100m^2 - 4.9.(1+m^2) = 0 \quad \text{(since } a = 1+m^2,$$
$$\Leftrightarrow 100m^2 - 36 - 36m^2 = 0 \quad b = 10m, c = 9)$$
$$\Leftrightarrow 64m^2 - 36 = 0$$
$$\Leftrightarrow m^2 = \tfrac{36}{64} = \tfrac{9}{16}$$
$$\Rightarrow m = \pm\tfrac{3}{4}$$

Hence the tangents from the point $(0, 5)$ have equations,
$$y = \tfrac{3}{4}x + 5 \quad \text{and} \quad y = -\tfrac{3}{4}x + 5$$
i.e. $\quad 4y = 3x + 20 \quad \text{and} \quad 4y = -3x + 20$
i.e. $\quad 4y - 3x = 20 \quad \text{and} \quad 4y + 3x = 20$

Example 4

Find the equation of the tangent at the point $A(1, -3)$ on the circle $x^2 + y^2 - 6x + 2y + 2 = 0$.

$$x^2 + y^2 - 6x + 2y + 2 = 0$$
$$\Leftrightarrow x^2 - 6x + 9 + y^2 + 2y + 1 = -2 + 9 + 1$$
$$\Leftrightarrow (x-2)^2 + (y+1)^2 = 8$$

hence the circle has centre $(2, -1)$ and radius $2\sqrt{2}$ units.

$A(1, -3)$ lies on the circle, $C(3, -1)$ is the centre of the circle and AC is perpendicular to the tangent at A.

Hence $\quad m_{AC} = \dfrac{-1+3}{3-1} = \dfrac{2}{2} = 1$

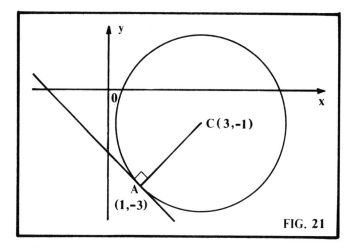

FIG. 21

and therefore the gradient of the tangent at A equals -1 and the tangent passes through $A(1, -3)$ and hence the tangent at A has equation $y + 3 = -1(x - 1)$,

i.e.
$$y + 3 = -x + 1$$
$$y + x + 2 = 0$$

Example 5

Calculate the length of the tangent from the point $(-2, 4)$ to the circle $x^2 + y^2 - 8x + 2y - 8 = 0$.

Comparing the given circle with
$$x^2 + y^2 + 2gx + 2fy + c = 0$$

$g = -4, f = 1$ and $c = -8$. Hence the centre of the circle is $(4, -1)$ and radius $= \sqrt{(16+1+8)} = 5$.

In figure 22 (next page),
$$CA = 5 \quad \text{(radius)}$$
$$PC^2 = (4+2)^2 + (-1-4)^2$$
$$= 36 + 25$$
$$= 61$$

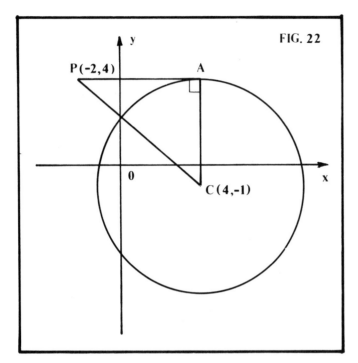

FIG. 22

Since the tangent PA and the radius CA make a right angle at the point of contact A,

$$PC^2 = PA^2 + AC^2$$
$$61 = PA^2 + 25 \Rightarrow PA^2 = 36$$
$$\Rightarrow PA = 6$$

hence the tangent from P has length 6 units.

ASSIGNMENT 3.4

1. Find the coordinates of the points of intersection of the given lines with the given circles:
   (i) $y - x + 2 = 0$ and $x^2 + y^2 = 10$
   (ii) $x - 2y - 1 = 0$ and $x^2 + y^2 = 2$
   (iii) $x - 3y = 1$ and $x^2 + y^2 = 5$
   (iv) $y - x - 1 = 0$ and $x^2 + y^2 + 2x + 2y + 1 = 0$
   (v) $2x - y - 4 = 0$ and $x^2 + y^2 + 4x - 4y - 17 = 0$
   (vi) $x + 3y - 7 = 0$ and $x^2 + y^2 - 4x - 1 = 0$

2. Show by finding the point of contact that the given lines are tangents to the given circles:
   (i) $x - y = 6$ and $x^2 + y^2 = 18$
   (ii) $x - 3y - 12 = 0$ and $x^2 + y^2 - 4x - 6 = 0$
   (iii) $3x + y - 2 = 0$ and $x^2 + y^2 + 6x - 2y = 0$
   (iv) $x - y = 10$ and $x^2 + y^2 - 4x + 8y + 12 = 0$
   (v) $x + 2y = 3$ and $x^2 + y^2 + 2y - 4 = 0$

3. Find the equation of the tangent from the given point to the given circle:
   (i) $(0, -5)$; $x^2 + y^2 = 5$
   (ii) $(-5, 0)$; $x^2 + y^2 = 5$
   (iii) $(0, 3)$; $x^2 + y^2 = 4$
   (iv) $(0, -4)$; $x^2 + y^2 = 8$
   (v) $(6, 0)$; $x^2 + y^2 = 12$

4. Verify that the given point lies on the given circle and find the equation of the tangent to the circle at the given point:
   (i) $x^2 + y^2 = 13$; $(2, -3)$
   (ii) $x^2 + y^2 = 26$; $(-1, -5)$
   (iii) $x^2 + y^2 = 20$; $(-4, 2)$
   (iv) $x^2 + y^2 - 2x - 10y + 17 = 0$; $(-2, 5)$
   (v) $x^2 + y^2 + 10x + 2y + 24 = 0$; $(-6, -2)$
   (vi) $x^2 + y^2 - 8x + 2y - 8 = 0$; $(1, 3)$
   (vii) $x^2 + y^2 - 10x - 2y - 11 = 0$; $(-1, 0)$

5. Calculate the length of the tangent from the given point to the given circle:
   (i) $(0, 5); x^2 + y^2 = 9$
   (ii) $(-13, 0); x^2 + y^2 = 25$
   (iii) $(4, 1); x^2 + y^2 + 12x - 8y + 27 = 0$
   (iv) $(4, -1); x^2 + y^2 + 6x - 8y + 1 = 0$
   (v) $(-2, 3); x^2 + y^2 - 10x + 3y + 15 = 0$
   (vi) $(2, 3); x^2 + y^2 - 4x + 10y + 17 = 0$

6. Show that the line $y = 2$ is a tangent to the circle $x^2 + y^2 - 4x - 4y + 8 = 0$.

7. Show that the line $x + 2y = 0$ is a tangent to the circle $x^2 + y^2 - 6x + 8y + 20 = 0$ and find the point of contact. The other tangent from the origin has equation $y = kx$. Find the value of k.

8. Find the equations of the tangents to the circle
$$x^2 + y^2 - 2x + 2y - 14 = 0$$
at their points of contact with the line $x - y + 2 = 0$.

9. Show that the tangent to the circle $x^2 + y^2 = 10$ at the point $(3, -1)$ is also a tangent to the circle
$$x^2 + y^2 + 5x + 5y - 10 = 0$$
and find the point of contact.

10. Find the value of c so that the line $x - 2y = c$ is a tangent to the circle $x^2 + y^2 = 20$.

11. Find the equation of the tangents at the points $(-1, 4)$ and $(-1, 0)$ on the circle $x^2 + y^2 + 6x - 4y + 5 = 0$ and show that they are perpendicular. Calculate the point of intersection of these tangents.

12. The line $2y + x = 12$ cuts the circle $x^2 + y^2 - 8x - 6y = 0$ at the points P and Q. Find the coordinates of P and Q and the equations of the tangents at P and Q.

If the tangents at P and Q intersect at R, show that the line RM where M is the mid-point of PQ passes through the centre of the circle.

ASSIGNMENT 3.5

SUPPLEMENTARY EXAMPLES

1. A is the point $(9, 0)$ and B the point $(1, 0)$. P is a point on the locus defined by $\{P(x, y): PA = 3PB\}$. Show that the locus of P is the circle with equation $x^2 + y^2 = 9$.
   Find the gradients of the two lines through the point $(0, 5)$ which touches this circle and find also the co-ordinates of the points of contact.

2. Prove that the circle $x^2 + y^2 + 2gx + 2fy + c = 0$ touches the x-axis $\Leftrightarrow g^2 = c$.
   Show that there are two circles which touch the x-axis and pass through the points $(-1, 2)$ and $(-1, 8)$ and find the equations of these two circles.

3. A circle passes through the points $P(-2, 1)$ and $Q(0, -1)$ and has its centre on the line $x + y + 1 = 0$.
   Find the equation of the circle.
   (*Hint*: Let the circle have equation
   $$x^2 + y^2 + 2gx + 2fy + c = 0$$
   and therefore its centre is $(-g, -f)$. Obtain three equations in g, f and c.)

4. Show that the equation of the circle which touches the y-axis at the origin may be written $x^2 + y^2 + kx = 0$.
   Find the equation of the circle which touches the y-axis at the point $(0, 0)$ and passes through the point $(5, 0)$. Write down the coordinates of the centre and the length of the radius of this circle.

5. A circle has centre $(-3, 4)$ and radius 5. Find its equation and the coordinates of the three points where the circle cuts the axes.
   Find the equations of the tangents at these three points and show that two of these tangents are parallel. If the third tangent cuts the two parallel ones at P and Q, find the coordinates of P and Q and hence calculate the lengths of the tangents from P and Q to the circle.

6. Show that the tangent at the point $A(-2, 1)$ on the circle $x^2 + y^2 = 5$ has equation $y - 2x = 5$. Show also that this line is a tangent to the circle
$$x^2 + y^2 - 6x - 2y - 10 = 0$$
and find its point of contact B with this circle.
   Calculate the length of the common tangent AB to these two circles.

7. Find the coordinates of the centre C and the length of the radius of the circle with equation
$$x^2 + y^2 - 6x - 91 = 0$$
   (i) The line with equation $2y = x + 7$ intersects this circle at the points A and B. Find the coordinates of A and B and write down the coordinates of M the mid-point of AB.
   (ii) If this circle is reflected in the line AB, find the equation of the image circle.

8. The line $y = kx$ intersects the circle with equation
$$x^2 + y^2 - 6x + 2y + 8 = 0$$
   (i) State the range of values of k for the line to intersect the circle in two distinct points.
   (ii) State the values of k for the line to be a tangent to the circle.

9. Find the coordinates of the points P and Q in which the line $y - 7x + 5 = 0$ intersects the circle $x^2 + y^2 = 5$.

(i) Find also the equations of the tangents to this circle which are parallel to PQ.
(ii) State the distance between these tangents.

10. As θ varies, the coordinates (x, y) of a variable point P are given by $x = 10 \cos \theta - 3$ and $y = 10 \sin \theta + 4$. Show that the locus of P is a circle and state the coordinates of the centre and length of the radius of the circle.
    The line $3y + 4x = 0$ intersects this circle at the points R and Q. Show that $\tan \theta = -4/3$ and hence or otherwise find the coordinates of R and Q.

11. The equation
$$x^2 + y^2 - 5y - 10 + k(y - 3x) = 0$$
represents a system of circles passing through the points of intersection of the line $y = 3x$ and the circle $x^2 + y^2 - 5y - 10 = 0$.
    Find,
    (i) the equation of the circle of this system which passes through the point $(-1, 3)$,
    (ii) the equation of the circle of this system whose centre is the point $(3, \frac{3}{2})$.

12. Show that the circle with centre (k, k) and touching the coordinate axes may be written
$$x^2 + y^2 - 2kx - 2ky + k^2 = 0$$
    (i) Find the equation of the two circles passing through the point $P(1, 2)$ and touching both axes.
    (ii) Write down the coordinates of Q the other point of intersection of these two circles.
    (iii) Find the coordinates of the point of intersection of the line PQ with the line of centres $C_1 C_2$, where C_1 and C_2 are the centres of the two circles and show that PQ divides $C_1 C_2$ in the ratio $1 : 7$.

13. Show that the condition for the line $y = mx$ to be

a tangent to the circle $x^2+y^2-2x+6y+k=0$ is $(9-k)m^2-6m+1-k=0$.

State the algebraic significance of the roots in m of this equation and hence prove that the two tangents from the origin to the circle will be perpendicular if and only if $k=5$.

14. If M(2, −1) is the mid-point of the chord PQ of the circle with equation $x^2+y^2-6x+8y-25=0$, show that the chord PQ has equation $x-3y=5$ and hence find the coordinates of P and Q.

    If T is the point (−2, 11), show that TP and TQ are tangents to the circle.

15. Show that the circle which touches the four lines $x=0$, $x=2$, $y=0$ and $y=2$ has equation
    $$x^2+y^2-2x-2y+1=0$$
    Show that the line $4x+3y=2$ is a tangent to this circle and find the coordinates of the point of contact.

16. Prove that the line $y=mx+c$ is a tangent to the circle
    $$x^2+y^2=a^2 \Leftrightarrow c^2=a^2(1+m^2)$$
    Show that the line $2y=3x+13$ is a tangent to the circle $x^2+y^2=13$ and find the coordinates of the point of contact.

17. A variable point P(x, y) distant d_1 from A(4, −2) and d_2 from B(−2, −2) satisfies the relation $d_1=2d_2$.
    (i) Show that the locus of P is the circle with equation $x^2+y^2+8x+4y+4=0$ and write down the co-ordinates of its centre.
    (ii) Find the equation of the tangent to this circle at the point $(-\frac{8}{5}, \frac{6}{5})$ on it.
    (iii) Show that this tangent also touches the circle
    $$x^2+y^2-4x+8y+16=0$$

ASSIGNMENT 3.6

Objective type items testing Sections 3.1–3.4.
Instructions for answering these items are given on page 16.

1. Which of the following equations does not represent a circle?
   A. $x^2+y^2-4x+2=0$
   B. $2x^2+2y^2-6x+2y=0$
   C. $x^2+y^2+y=0$
   D. $x^2+2y^2+4y-x=0$
   E. $x^2+y^2-7=0$

2. The circle $2x^2+2y^2+4x-3y+1=0$ has as centre the point
   A. $(-2, \frac{3}{2})$
   B. $(1, -\frac{3}{4})$
   C. $(-4, 3)$
   D. $(-2, \frac{3}{2})$
   E. $(-1, \frac{3}{4})$

3. The circle with centre $(\frac{1}{2}, -\frac{3}{2})$ and radius 5 has equation
   A. $2x^2+2y^2-4x+6y+45=0$
   B. $x^2+y^2-2x+3y+2\frac{1}{2}=0$
   C. $4x^2+4y^2-8x+12y+9=0$
   D. $x^2+y^2-2x+3y-24=0$
   E. $2x^2+2y^2-4x+6y-45=0$

4. The equation of the circle obtained by reflecting the circle $x^2+y^2-6x+5y-12=0$ in the origin is
   A. $x^2+y^2-5x+6y-12=0$
   B. $x^2+y^2-6x-5y-12=0$
   C. $x^2+y^2+6x+5y-12=0$
   D. $x^2+y^2+6x-5y-12=0$
   E. $x^2+y^2+5x-6y-12=0$

5. Which of the following statements is false for the circle $x^2+y^2-10y=0$?
   A. The circle passes through the origin.
   B. The circle has radius 5 units.
   C. The circle touches the x-axis.
   D. The circle has its centre on the y-axis.
   E. The circle passes through the point (10, 10).

6. The circle obtained by rotating the circle with equation $x^2+y^2-7x+9y-14=0$ through an angle of 90° anti-clockwise about the origin has equation
   A. $x^2+y^2-7x-9y-14=0$
   B. $x^2+y^2+9x-7y+14=0$
   C. $x^2+y^2-7x+9y+14=0$
   D. $x^2+y^2-9x-7y-14=0$
   E. $x^2-y^2+9x-7y+14=0$

7. The tangent at the point (1, 5) on the circle
   $$x^2+y^2-x-3y-10=0$$
   has equation
   A. $7y+x=36$
   B. $y-7x+2=0$
   C. $7y-x=34$
   D. $y+7x=12$
   E. none of these.

8. If the locus of the point P(x, y) is defined by
   $$\{P(x,y): PA = PB\}$$
   where A is (3, 0) and B is (−3, 0), then the locus of P
   A. is a circle
   B. is a straight line
   C. is some other curve
   D. has equation $x^2+y^2=9$
   E. has equation $y^2+12x=0$

9. The circle with equation $x^2+y^2+10x+6y+9=0$,
   (1) touches the y-axis
   (2) cuts the x-axis in two distinct points
   (3) passes through the origin

10. P(x, y) is a member of the set of points satisfying the equation $x^2+y^2=4$. M is the mid-point of OP.
    (1) the locus of M has equation $x^2+y^2=1$
    (2) the coordinates of M are $\left(\dfrac{x}{2}, \dfrac{y}{2}\right)$
    (3) the length of OM is 1 unit

11. (1) The circle with equation $x^2+y^2+2gx+2fy+c=0$ touches the x-axis at the point (0, 0).
    (2) $g = c = 0, f \neq 0$.

12. (1) The line $y = mx+c$ is a tangent to the circle
    $$x^2+y^2 = c^2, \quad c \neq 0$$
    (2) $m = 0$.

TRIGONOMETRY

UNIT 1: REVISION

TRIGONOMETRICAL FUNCTIONS FOR ANY ANGLE $a°$ FROM 0° TO 360°

1.1

Let (x, y) be the coordinates of any point P, such that angle XOP = $a°$.

(i) *When P(x, y) is in the 1st quadrant*
i.e. when $0 < a < 90$ (figure 1)

$$\sin a° = \frac{y}{r} = \frac{+ve}{+ve} = +ve$$

$$\cos a° = \frac{x}{r} = \frac{+ve}{+ve} = +ve$$

$$\tan a° = \frac{y}{x} = \frac{+ve}{+ve} = +ve$$

(ii) *When P(x, y) is in the 2nd quadrant*
i.e. when $90 < a < 180$ (figure 2)

$$\sin a° = \frac{y}{r} = \frac{+ve}{+ve} = +ve$$

$$\cos a° = \frac{x}{r} = \frac{-ve}{+ve} = -ve$$

$$\tan a° = \frac{y}{x} = \frac{+ve}{-ve} = -ve$$

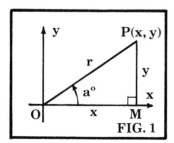
FIG. 1

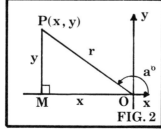
FIG. 2

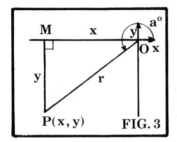
FIG. 3

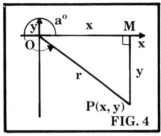
FIG. 4

(iii) *When $P(x, y)$ is in the 3rd quadrant*
i.e. when $180 < a < 270$ (figure 3)

$$\sin a° = \frac{y}{r} = \frac{-\text{ve}}{+\text{ve}} = -\text{ve}$$

$$\cos a° = \frac{x}{r} = \frac{-\text{ve}}{+\text{ve}} = -\text{ve}$$

$$\tan a° = \frac{y}{x} = \frac{-\text{ve}}{-\text{ve}} = +\text{ve}$$

(iv) *When $P(x, y)$ is in the 4th quadrant*
i.e. when $270 < a < 360$ (figure 4)

$$\sin a° = \frac{y}{r} = \frac{-\text{ve}}{+\text{ve}} = -\text{ve}$$

$$\cos a° = \frac{x}{r} = \frac{+\text{ve}}{+\text{ve}} = +\text{ve}$$

$$\tan a° = \frac{y}{x} = \frac{-\text{ve}}{+\text{ve}} = -\text{ve}$$

These four illustrations give us the sign of the trigonometrical functions for any angle between 0° and 360°. For angles greater than 360°, for example 500° we note that $500° = 360° + 140°$. Thus the point $P(x, y)$ is situated in the second quadrant. Therefore the sign of each trigonometrical function for 500° is the same as the sign of the trigonometrical function for 140°.

i.e. $\quad\sin 500° = \sin 140°$ and is positive
$\cos 500° = \cos 140°$ and is negative
$\tan 500° = \tan 140°$ and is negative

Thus between 360° and 720° we have a repetition of the *sign* changes as we have from 0° to 360° and so on for each additional 360°. So sin 280° is negative (since 280° is in the fourth quadrant).

$$\text{Cos } 440° = \cos (360° + 80°)$$

and is positive since 80° is in the first quadrant.

$$\text{Tan } 580° = \tan (360° + 220°)$$

and is positive since 220° is in the third quadrant.

The *sign* of the trigonometrical functions may be remembered from figure 5 which indicates the quadrants in which each function is positive.

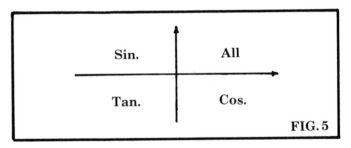
FIG. 5

TRIGONOMETRICAL FUNCTIONS FOR CERTAIN RELATED ANGLES—PART (i)

TRIGONOMETRICAL FUNCTIONS FOR $(90-a)°$—COMPLEMENTARY ANGLES (FIGURE 6)

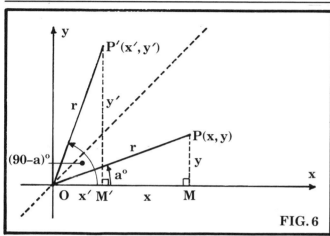

FIG. 6

Let $P(x, y)$ be any point such that $\angle XOP = a°$, $a < 90$.
Under reflection in the line $y = x$, $P(x, y) \to P'(x', y')$ such that $y' = x$, $x' = y$, $OP' = OP = r$ and $\angle XOP' = (90-a)°$

$$\sin(90-a)° = \frac{y'}{r} = \frac{x}{r} = \cos a°$$

$$\cos(90-a)° = \frac{x'}{r} = \frac{y}{r} = \sin a°$$

Hence, the sine of an angle = the cosine of its complement and the cosine of an angle = the sine of its complement.

Example 1
$$\sin 60° = \cos(90-60)° = \cos 30°$$
$$\cos 52° = \cos(90-52)° = \sin 38°$$

TRIGONOMETRICAL FUNCTIONS FOR $(180-a)°$—SUPPLEMENTARY ANGLES (FIGURE 7)

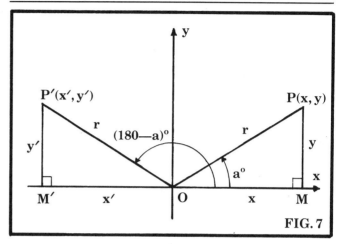

FIG. 7

Let $P(x, y)$ be any point such that $\angle XOP = a°$, $0 < a < 180$.
Under reflection in the y-axis $P(x, y) \to P'(x', y')$, such that $x' = -x$, $y' = y$, $OP' = OP = r$ and $\angle XOP' = (180-a)°$

$$\sin(180-a)° = \frac{y'}{r} = \frac{y}{r} = \sin a°$$

$$\cos(180-a)° = \frac{x'}{r} = \frac{-x}{r} = -\frac{x}{r} = -\cos a°$$

$$\tan(180-a)° = \frac{y'}{x'} = \frac{y}{-x} = -\frac{y}{x} = -\tan a°$$

hence,

the sine of an angle = the sine of its supplement
the cosine of an angle = − the cosine of its supplement
the tangent of an angle = − the tangent of its supplement.

Example 2

$$\sin 120° = \sin(180-120)° = \sin 60°$$
$$\cos 108° = -\cos(180-108)° = -\cos 72°$$
$$\tan 142° = -\tan(180-142)° = -\tan 38°$$

Example 3

If $\sin x° = \frac{8}{17}$ and $90 < x < 180$, find without tables the value of $\cos x°$ and $\tan x°$.

Draw a right-angled triangle POM, with exterior angle PON = $x°$ (figure 8).

$$\angle POM = (180-x)° \text{ and } \sin x° = \tfrac{8}{17}$$

Hence PM = 8 and OP = 17 since $\sin x° = \sin(180-x)°$. By Pythagoras the third side of triangle POM, namely OM is equal to 15.

$$\therefore \cos(180-x)° = \tfrac{15}{17} \text{ and } \tan(180-x)° = \tfrac{8}{15}$$

But $90 < x < 180$ and both $\cos x°$ and $\tan x°$ are negative in this range.

Hence $\cos x° = -\frac{15}{17}$ and $\tan x° = -\frac{8}{15}$.

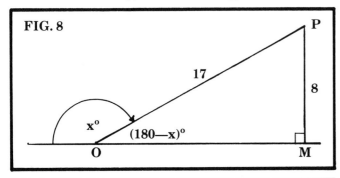

TRIGONOMETRICAL FUNCTIONS OF 30°, 45°, 60°

For 60° and 30°

△ABC is equilateral of side two units. M is the mid-point of BC (figure 9).

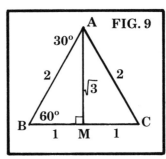
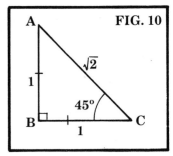

$\therefore$ AM is perpendicular to BC

$\therefore$ BM = MC = 1 unit

By Pythagoras

$$AB^2 = BM^2 + AM^2$$
$$4 = 1 + AM^2$$
$$AM^2 = 3 \Rightarrow AM = \sqrt{3}$$

$$\therefore \sin 60° = \frac{AM}{AB} = \frac{\sqrt{3}}{2} \qquad \sin 30° = \frac{BM}{AB} = \frac{1}{2}$$

$$\cos 60° = \frac{BM}{AB} = \frac{1}{2} \qquad \cos 30° = \frac{AM}{AB} = \frac{\sqrt{3}}{2}$$

$$\tan 60° = \frac{AM}{BM} = \frac{\sqrt{3}}{1} \qquad \tan 30° = \frac{BM}{AM} = \frac{1}{\sqrt{3}}$$

For 45°

△ABC is an isosceles right-angled triangle with AB = AC = 1 unit (figure 10).

$$\therefore BC^2 = AB^2 + AC^2 = 1 + 1 = 2$$
$$\Rightarrow BC = \sqrt{2}$$
$$\angle ACB = \angle ABC = 45°$$
$$\sin 45° = \frac{AC}{BC} = \frac{1}{\sqrt{2}}$$

$$\cos 45° = \frac{AB}{BC} = \frac{1}{\sqrt{2}}$$

$$\tan 45° = \frac{AC}{AB} = 1$$

Hence table giving exact values of trigonometric functions for certain angles.

$x°$	0°	30°	45°	60°	90°	120°	135°	150°	180°
$\sin x°$	0	$\frac{1}{2}$	$\frac{1}{\sqrt{2}}$	$\frac{\sqrt{3}}{2}$	1	$\frac{\sqrt{3}}{2}$	$\frac{1}{\sqrt{2}}$	$\frac{1}{2}$	0
$\cos x°$	1	$\frac{\sqrt{3}}{2}$	$\frac{1}{\sqrt{2}}$	$\frac{1}{2}$	0	$-\frac{1}{2}$	$-\frac{1}{\sqrt{2}}$	$-\frac{\sqrt{3}}{2}$	-1
$\tan x°$	0	$\frac{1}{\sqrt{3}}$	1	$\frac{\sqrt{3}}{2}$		$-\sqrt{3}$	-1	$-\frac{1}{\sqrt{3}}$	0

ASSIGNMENT 1.1

1. State the sign (positive or negative) of each of the following:
   (i) sin 240°   (ii) cos 120°   (iii) tan 140°
   (iv) tan 330°   (v) sin 382°   (vi) tan 480°
   (vii) cos 310° (viii) cos 760° (ix) sin 890°
   (x) tan 260°   (xi) cos 540°   (xii) sin 823°

2. State the algebraic sign of the sine, cosine and tangent of the angle $a°$, if,
   (a) $180 < a < 270$ (b) $360 < a < 450$
   (c) $630 < a < 720$ (d) $450 < a < 540$

3. If $\cos x°$ is positive, and $x < 360$, state the possible limits between which x may lie.

4. If $\sin x°$ is negative, and $x < 360$, state the possible limits between which x may lie.

5. State the quadrant or quadrants in which the angle x may lie, if,
   (i) $\sin x°$ and $\tan x°$ have the same sign
   (ii) $\cos x° = 0·423$ and $\tan x° = -2·145$
   (iii) $\sin x°$ and $\cos x°$ have both negative signs
   (iv) $\tan x°$ is positive but $\cos x°$ is negative

6. Use tables to find the value of,
   (i) sin 142°   (ii) tan 98°   (iii) cos 163°
   (iv) cos 170° (v) sin 136°   (vi) tan 100°
   (vii) cos 180° (viii) tan 180°

7. Give the *exact* value of the following:
   (i) cos 30°   (ii) sin 135°   (iii) tan 60°
   (iv) sin 45°   (v) tan 135°   (vi) cos 180°
   (vii) cos 150° (viii) tan 30°

8. Give the solution set of the following equations, for $0 \leqq x \leqq 180$.
   (i) $\tan x° = 1$ (ii) $\cos x° = -\frac{1}{2}$
   (iii) $\sin x° = \frac{\sqrt{3}}{2}$ (iv) $\sin x° = \sin 150°$
   (v) $\sin x° = -\frac{1}{\sqrt{2}}$ (vi) $\tan x° = -\sqrt{3}$
   (vii) $\cos x° = -0·339$

9. Use tables to find the value of,
   (i) cos 162°   (ii) tan 143°   (iii) sin 153°
   (iv) cos 135° (and compare with the exact value of cos 135°)

(v) sin 120° (and compare with the exact value of sin 120°)
(vi) tan 150° (and compare with the exact value of tan 150°)

1.3

10. Simplify:
    (i) $\cos(180-a)°$
    (ii) $\tan(180-p)°$
    (iii) $\sin(90-x)°$
    (iv) $\sin(180-q)°$
    (v) $\cos(90-t)°$

11. Write the following in terms of an acute angle,
    (i) cos 170° (ii) sin 118° (iii) tan 92°

12. If $\sin x° = \cos x°$, find the value of x, for $0 \leq x \leq 180$.

13. If $\cos x° = \frac{3}{5}$, find the exact value of $\sin x°$ for $0 < x < 90$.

14. If $\tan x° = -\frac{5}{12}$ and $90 < x < 180$, write down the *exact* value of $\sin x°$ and $\cos x°$.

15. Solve the following equations:
    (i) $\sin x° + \sin 42° = 0$ $(0 \leq x \leq 180)$
    (ii) $\cos x° + \cos 70° = 0$ $(0 \leq x \leq 180)$
    (iii) $\tan x° = -\sin 38°$ $(0 \leq x \leq 180)$
    (iv) $\cos x° = -\sin 135°$ $(0 \leq x \leq 180)$

16. By using the exact values of the angles, verify that
    (i) $\sin^2 30° + \cos^2 30° = 1$
    (ii) $\sin^2 90° + \cos^2 90° = 1$
    (iii) $\dfrac{\sin 60°}{\cos 60°} = \tan 60°$
    (iv) $\tan 60° = \dfrac{1}{\tan 30°}$
    (v) $\cos^2 120° + \sin^2 120° = 1$
    (vi) $2 \sin 45° \cos 45° = \sin 90°$
    (vii) $\cos 60° = 2 \cos^2 30° - 1$

17. If $\cos 60° = \frac{1}{2}$, without using tables write down the values of cos 120°, sin 30° and sin 150°.

TRIGONOMETRICAL FUNCTIONS FOR CERTAIN RELATED ANGLES—PART (ii)

TRIGONOMETRICAL FUNCTIONS FOR $(180 + a)°$ (FIGURE 11)

Let $P(x, y)$ be any point such that $\angle XOP = a°$, $0 < a < 90$.

Under an anticlockwise rotation of 180°, $P(x, y) \rightarrow P'(x', y')$ such that $x' = -x$, $y' = -y$, $OP' = OP = r$ and $\angle XOP' = (180 + a)°$

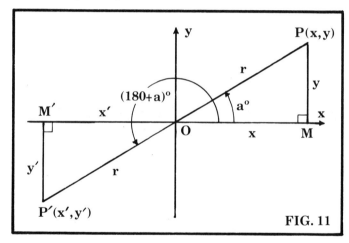

FIG. 11

$$\sin(180+a)° = \frac{y'}{r} = \frac{-y}{r} = -\frac{y}{r} = -\sin a°$$

$$\cos(180+a)° = \frac{x'}{r} = \frac{-x}{r} = -\frac{x}{r} = -\cos a°$$

$$\tan(180+a)° = \frac{y'}{x'} = \frac{-y}{-x} = \frac{y}{x} = \tan a°$$

Example 1

$$\sin 240° = \sin(180+60)° = -\sin 60° = -\frac{\sqrt{3}}{2}$$

$$\cos 225° = \cos(180+45)° = -\cos 45° = -\frac{1}{\sqrt{2}}$$

$$\tan 210° = \tan(180+30)° = \tan 30° = \frac{1}{\sqrt{3}}$$

TRIGONOMETRICAL FUNCTIONS FOR $(360-a)°$ (FIGURE 12)

Let $P(x, y)$ be any point such that $\angle XOP = a°$, $0 < a < 90$.
Under a reflection in the x-axis $P(x, y) \to P'(x', y')$ such that $x' = x$, $y' = -y$, $OP' = OP = r$ and the reflex $\angle XOP' = (360-a)°$

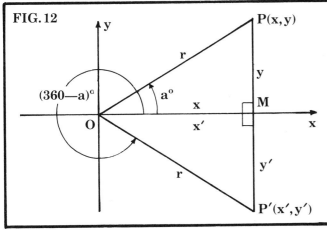

FIG. 12

$$\sin(360-a)° = \frac{y'}{r} = \frac{-y}{r} = -\frac{y}{r} = -\sin a°$$

$$\cos(360-a)° = \frac{x'}{r} = \frac{x}{r} = \cos a°$$

$$\tan(360-a)° = \frac{y'}{x'} = \frac{-y}{x} = -\frac{y}{x} = -\tan a°$$

Example 2

$$\sin 330° = \sin(360-30)° = -\sin 30° = -\tfrac{1}{2}$$

$$\cos 300° = \cos(360-60)° = \cos 60° = \tfrac{1}{2}$$

$$\tan 315° = \tan(360-45)° = -\tan 45° = -1$$

Example 3

Find the solution sets of the following equations

(a) $\cos x° = \tfrac{1}{2}$ $0 \leq x \leq 360$

(b) $\tan x° = 1$ $0 \leq x \leq 360$

(c) $\sin x° = -\dfrac{\sqrt{3}}{2}$ $0 \leq x \leq 360$

(a) $\cos x° = \tfrac{1}{2} \Rightarrow x = 60$ for the acute angle x, but $\cos x°$ is also positive for $270 < x < 360$, hence $\cos x° = \tfrac{1}{2}$ has solution set $\{60, 300\}$.

(b) $\tan x° = 1 \Rightarrow x = 45$ for the acute angle x, but $\tan x°$ is also positive for $180 < x < 270$. Hence $\tan x° = 1$ has solution set $\{45, 225\}$.

(c) $\sin x° = \dfrac{\sqrt{3}}{2} \Rightarrow x = 60°$ for the actue angle x, but $\sin x° = -\dfrac{\sqrt{3}}{2}$ and $\sin x°$ is negative for $180 < x < 270$ and for $270 < x < 360$. Hence $\sin x° = -\dfrac{\sqrt{3}}{2}$ has solution set $\{240, 300\}$.

TRIGONOMETRICAL FUNCTIONS FOR $(-a°)$

Let $P(x, y)$ be any point such that $\angle XOP = a°$, $0 < a < 90$.
Under a reflection in the x-axis $P(x, y) \to P'(x', y')$ such that $x' = x$, $y' = -y$, $OP' = OP = r$ and the acute $\angle XOP' = -a°$ (see figure 13, over page).

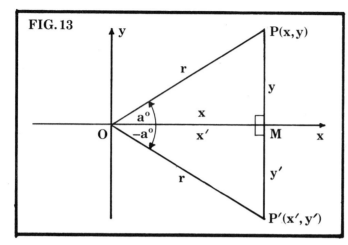

FIG. 13

$$\sin(-a°) = \frac{y'}{r} = \frac{-y}{r} = -\frac{y}{r} = -\sin a°$$

$$\cos(-a°) = \frac{x'}{r} = \frac{x}{r} = \cos a°$$

$$\tan(-a°) = \frac{y'}{x'} = \frac{-y}{x} = -\frac{y}{x} = -\tan a°$$

This also holds if $a > 90$ and gives us the trigonometrical functions of a negative angle in terms of the trigonometrical functions of the corresponding positive angle.

Example 4

$$\sin(-40°) = -\sin 40° = -0.643$$
$$\cos(-120°) = \cos 120° = -\cos 60° = -\tfrac{1}{2}$$
$$\tan(-210°) = -\tan 210° = -\tan 30° = -\frac{1}{\sqrt{3}}$$

Again

$$\sin(-330°) = -\sin 330° = -(-\sin 30°) = \frac{\sqrt{3}}{2}$$

ASSIGNMENT 1.2

1. Write down the values of the following. Where possible give the exact value, for example,

   $$\cos 210° = -\cos 30° = -\frac{\sqrt{3}}{2} \text{ exact value}$$

   but

   $$\cos 228° = -\cos 48° = -0.699 \text{ from tables}$$

   (i) $\cos 110°$ (ii) $\sin 225°$ (iii) $\tan 300°$
   (iv) $\sin 345°$ (v) $\tan 230°$ (vi) $\cos 315°$
   (vii) $\sin 140°$ (viii) $\cos 240°$ (ix) $\sin 131°$
   (x) $\tan 136°$ (xi) $\sin 270°$ (xii) $\cos 360°$

2. Express each of the following in terms of an acute angle with the appropriate sign, (example: $\cos 150° = -\cos 30°$)
   (i) $\tan 215°$ (ii) $\sin 346°$ (iii) $\cos 272°$
   (iv) $\sin 192°$ (v) $\cos 269°$ (vi) $\tan 310°$

3. Express the following in terms of an acute angle with appropriate sign.
   (i) $\sin(-50°)$ (ii) $\sin(-120°)$
   (iii) $\tan(160°)$ (iv) $\cos(-300°)$
   (v) $\cos(-118°)$ (vi) $\tan(-210°)$

4. Write down the values of the following (where possible give the exact value).
   (i) $\cos(-225°)$ (ii) $\sin(-300°)$
   (iii) $\tan(-42°)$ (iv) $\sin(-200°)$
   (v) $\tan(-240°)$ (vi) $\cos(-10°)$
   (vii) $\sin(-270°)$ (viii) $\cos(-180°)$
   (ix) $\cos(-315°)$

5. If $\tan a° = \frac{20}{21}$ and $180 < a < 270$, find the exact value of $\sin a°$ and $\cos a°$.

6. If $\cos x° = \frac{12}{13}$ and $270 < x < 360$, find the exact value of $\sin x°$ and $\tan x°$.

7. Find the solution set of the following equations for $0 \leq x \leq 360$.

   (i) $\cos x° = \frac{1}{2}$ (ii) $\tan x° = \frac{1}{\sqrt{3}}$ (iii) $\cos x° = -\frac{1}{\sqrt{2}}$

   (iv) $\sin x° = -\frac{1}{2}$ (v) $\tan x° = -1$

8. By using the exact values of the angles, verify the following:

   (i) $\cos^2 315° + \sin^2 315° = 1$

   (ii) $\dfrac{\sin 210°}{\cos 210°} = \tan 210°$

   (iii) $\dfrac{\sin 330°}{\cos 330°} = -\tan 30°$

9. Find the solution set of the equations for $0 \leq x \leq 360°$.

   (i) $\sin x° = -0·759$ (ii) $\sin x° + \cos x° = 0$

   (iii) $\cos x° = 0·895$ (iv) $\tan x° = 1·253$

TRIGONOMETRICAL RELATIONS

In figure 14 angle $a°$ is any angle. Consider the case when $a°$ is an obtuse angle

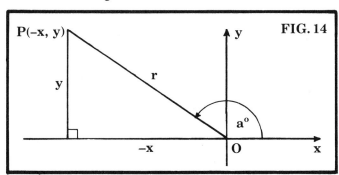

FIG. 14

$\therefore \angle XOP = a°$ and P is the point $(-x, y)$

hence

$$\sin a° = \frac{y}{r} \text{ and } \cos a° = \frac{-x}{r}$$

(i) $\sin^2 a° + \cos^2 a° = \left(\dfrac{y}{r}\right)^2 + \left(\dfrac{-x}{r}\right)^2$

$= \dfrac{y^2}{r^2} + \dfrac{x^2}{r^2}$

$= \dfrac{y^2 + x^2}{r^2} = \dfrac{r^2}{r^2}$

(since $y^2 + x^2 = r^2$ by Pythagoras)

$= 1.$

Hence for any angle A:

$$\sin^2 A + \cos^2 A = 1$$

Note: that we also have $\cos^2 A = 1 - \sin^2 A$ and $\sin^2 A = 1 - \cos^2 A$.

(ii)

$$\frac{\sin a°}{\cos a°} = \frac{y/r}{-x/r} = \frac{y}{r} \div \left(\frac{-x}{r}\right) = \frac{y}{r} \times \frac{r}{-x}$$

$= \dfrac{y}{-x} = \tan a°$

Hence for any angle A:

$$\frac{\sin A}{\cos A} = \tan A$$

Note that $\sin A = \cos A \cdot \tan A$.

These two trigonometrical relations may be used in proving certain identities or in elimination.

1.4

Example 1

Prove that

$$\frac{1}{1+\cos \alpha} + \frac{1}{1-\cos \alpha} = \frac{2}{\sin^2 \alpha}$$

We prove this by taking the left hand side (L.H.S.) and by using relations (i) and/or (ii) reduce it to the right hand side (R.H.S.).

Hence:

$$\text{L.H.S.} = \frac{1}{1+\cos \alpha} + \frac{1}{1-\cos \alpha} = \frac{1-\cos \alpha + 1 + \cos \alpha}{(1+\cos \alpha)(1-\cos \alpha)}$$

$$= \frac{2}{1-\cos^2 \alpha}$$

$$= \frac{2}{\sin^2 \alpha}$$

(since $\sin^2 \alpha + \cos^2 \alpha = 1 \Rightarrow \sin^2 \alpha = 1 - \cos^2 \alpha$).

Example 2

Prove that

$$\frac{1-\tan^2 A}{1+\tan^2 A} = 2\cos^2 A - 1$$

$$\text{L.H.S.} = \frac{1-\tan^2 A}{1+\tan^2 A} = \frac{1 - \dfrac{\sin^2 A}{\cos^2 A}}{1 + \dfrac{\sin^2 A}{\cos^2 A}}$$

$$= \frac{\dfrac{\cos^2 A - \sin^2 A}{\cos^2 A}}{\dfrac{\cos^2 A + \sin^2 A}{\cos^2 A}}$$

$$= \frac{\cos^2 A - \sin^2 A}{\cos^2 A} \div \frac{1}{\cos^2 A}$$

(since $\cos^2 A + \sin^2 A = 1$)

$$= \frac{\cos^2 A - \sin^2 A}{\cos^2 A} \times \frac{\cos^2 A}{1}$$

$$= \cos^2 A - \sin^2 A$$

$$= \cos^2 A - (1 - \cos^2 A) \quad \text{(since } \sin^2 A = 1 - \cos^2 A\text{)}$$

$$= 2\cos^2 A - 1$$

Example 3

If we are given two equations containing one variable then the process of deriving another equation *not* containing the variable is called **elimination**.

Eliminate θ from the equations, $x = r \cos \theta$, $y = r \sin \theta$.

$$x = r \cos \theta \Rightarrow \cos \theta = \frac{x}{r}$$

$$y = r \sin \theta \Rightarrow \sin \theta = \frac{y}{r}$$

but

$$\cos^2 \theta + \sin^2 \theta = 1 \Rightarrow \frac{x^2}{r^2} + \frac{y^2}{r^2} = 1$$

$$\Rightarrow x^2 + y^2 = r^2$$

This equation is called the **Eliminant**.

Example 4

Eliminate x from the equations,

$$2 \cos x + \sin x = a \quad \ldots \ldots \ldots \ldots (1)$$

$$\cos x - \sin x = b \quad \ldots \ldots \ldots \ldots (2)$$

Add equations (1) and (2),

$$3\cos x = a+b$$
$$\Rightarrow \cos x = \tfrac{1}{3}(a+b) \quad \ldots\ldots\ldots\ldots (3)$$

from equation (2),

$$\cos x - \sin x = b$$
$$\Rightarrow \sin x = \cos x - b$$
$$= \tfrac{1}{3}(a+b) - b \quad \text{using equation (3)}$$
$$= \tfrac{1}{3}(a+b-3b)$$
$$= \tfrac{1}{3}(a-2b)$$

but

$$\sin^2 x + \cos^2 x = 1 \Rightarrow \frac{(a-2b)^2}{9} + \frac{(a+b)^2}{9} = 1$$
$$\Rightarrow (a-2b)^2 + (a+b)^2 = 9$$
$$\Rightarrow a^2 - 4ab + 4b^2 + a^2 + 2ab + b^2 = 9$$
$$\Rightarrow 2a^2 - 2ab + 5b^2 = 9$$

ASSIGNMENT 1.3

1. Express each of the following in terms of a single angle.
   (i) $\dfrac{\sin 25°}{\cos 25°}$
   (ii) $\dfrac{\sin 50°}{\sin 40°}$
   (iii) $\tan A \times \cos A$
   (iv) $(1 - \cos^2 A)^{\frac{1}{2}}$

2. Simplify:
   (i) $\cos^2 a° + \cos^2(90-a)°$
   (ii) $\sin^2 a° + \cos^2(180-a)°$

3. Find the value of $\sin^2 145° + \cos^2 145°$.

4. Given $\tan P = \tfrac{3}{4}$ and angle P acute, evaluate $\dfrac{\sin P - \cos P}{\sin P + \cos P}$ without using tables.

5. If $x = a\cos x°$ and $y = b\sin x°$, prove that $\dfrac{x^2}{a^2} + \dfrac{y^2}{b^2} = 1$.

6. Express:
   (i) $1 + 3\sin^2 A$ in terms of $\cos^2 A$
   (ii) $1 - 4\cos^2 A$ in terms of $\sin^2 A$

Prove that,

7. $\sin A \cdot \tan A = \dfrac{1 - \cos^2 A}{\cos A}$

8. $(\sin x° + \cos x°)^2 = 1 + 2\cos x° \sin x°$

9. $\dfrac{1}{1 - \sin \alpha} + \dfrac{1}{1 + \sin \alpha} = \dfrac{2}{\cos^2 \alpha}$

10. $1 + 3\tan^2 \theta = \dfrac{1 + 2\sin^2 \theta}{\cos^2 \theta}$

11. $\dfrac{\cos^2 A}{1 + \sin A} = 1 - \sin A$

12. $\dfrac{2\tan B}{1 + \tan^2 B} = 2\sin B \cos B$

13. $\dfrac{\sin A}{1 + \cos A} + \dfrac{1 + \cos A}{\sin A} = \dfrac{2}{\sin A}$

14. Eliminate the variable θ or α from each of the following pairs of equations.
    (i) $x = 4\cos\theta,\ y = 4\sin\theta$
    (ii) $x = a\cos\theta,\ y = b\sin\theta$
    (iii) $\cos\alpha - 2\sin\alpha = m$, and $\sin\alpha + 2\cos\alpha = n$
    (iv) $\sin\alpha + \cos\alpha = a$ and $\sin\alpha - \cos\alpha = b$

15. Eliminate θ from the equations,
$$x\cos^2\theta = 1 + \sin\theta$$
$$y = 1 - \sin\theta$$

1.5 GRAPHS OF THE TRIGONOMETRICAL FUNCTIONS

THE SINE FUNCTION $f: x° \to \sin x°$

Since the sine function has a period of 360°, the graph will repeat itself after 360°.

To draw the graph we construct a table of values for $\sin x°$ for certain values of x between 0 and 360. When the table has been filled in for values of x up to 90 the remaining numerical values are simply a repetition of those already obtained with the appropriate sign.

x	0	30	45	60	90	120	135	150	180
$\sin x°$	0	0·5	0·71	0·87	1	0·87	0·71	0·5	0

x	210	225	240	270	300	315	330	360
$\sin x°$	−0·5	−0·71	−0·87	−1	−0·87	−0·71	−0·5	0

THE COSINE FUNCTION $f: x° \to \cos x°$

Since the cosine function has a period of 360°, the graph will repeat itself after 360°.

We construct a table similar to the table for the sine function.

x	0	30	45	60	90	120	135	150	180
$\cos x°$	1	0·87	0·71	0·5	0	−0·5	−0·71	−0·87	−1

x	210	225	240	270	300	315	330	360
$\cos x°$	−0·87	−0·71	−0·5	0	0·5	0·71	0·87	1

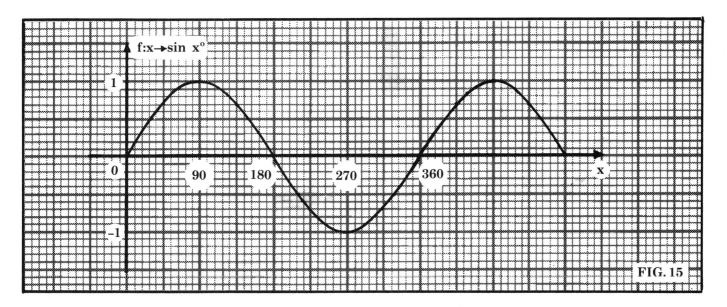

FIG. 15

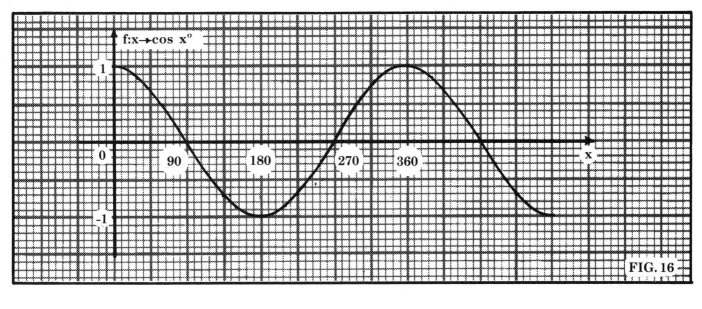

FIG. 16

THE TANGENT FUNCTION $f: x° \to \tan x°$

Since the tangent function has a period of 180°, the graph will repeat itself after 180°.

We construct a similar table as for the sine and cosine functions.

x	0	30	45	60	90	120	135	150	180
$\tan x°$	0	0·58	1	1·74		−1·74	−1	−0·58	0

x	210	225	240	270	300	315	330	360
$\tan x°$	0·58	1	1·74		−1·74	−1	−0·58	0

Note:

1. $\tan x°$ is *positive and increases* in the interval $[0, 90)$ i.e. the interval is closed at $0°$ since $\tan 0° = 0$ and the interval is open at $90°$ since $\tan 90°$ has no real value.
2. $\tan x°$ is *negative and increases* in the interval $(90, 180]$ i.e. the interval is open at $90°$ for $\tan 90°$ has no real value and the interval is closed at $180°$ for $\tan 180° = 0$.
3. *From tables*, since $\tan 0° = 0$ and $\tan 89·9° = 573·0$, $\tan x°$ is *positive and increases* as $x°$ increases from $0°$ to $89·9°$.
   Again, since
   $$\tan 90·1° = -\tan(180 - 90·1)° = -\tan 89·9°$$
   then
   $$\tan 90·1° = -573·0 \text{ and } \tan 180° = 0$$
   hence $\tan x°$ is *negative and increases* as $x°$ increases from $90·1°$ to $180°$.
4. A similar situation occurs at 270°.

Hence graph of $f: x° \to \tan x°$ (figure 17 on opposite page).

Periodicity

As has already been stated the sine, cosine and tangent functions are periodic functions. Since the graphs of these functions fit on to themselves after a translation the graphs are called **periodic** graphs.

The part of the graph of the sine function between $x = 0$ and $x = 360$ repeats itself after every 360°. It therefore has a period of 360°.

Also since the part of the graph between $x = 0$ and $x = 720$ repeats itself after every 720°, it has a period of 720°.

Similar repetitions of the graph occur after every multiple of 360° and give a period of $360n°$ where n is an integer.

We therefore say that the "least" period of the sine function is 360°.

Similarly the least period of the cosine function is 360° and the least period of the tangent function is 180°.

Example 1

State the least period of

    (i) $\cos x°$ (ii) $\cos 2x°$ (iii) $\cos^2 x°$

(i) $\cos 0° = 1$
$\cos 180° = -1$
$\cos 360° = 1$
hence the least period of $\cos x°$ is 360°.

(ii) $\cos 2x° = 1 \Rightarrow 2x = 0, 360, 720$
$\Rightarrow x = 0, 180, 360$
hence the least period of $\cos 2x°$ is 180°.

(iii) *Note:* $0 \leq \cos^2 x° \leq 1$ for all values of x
and $\cos^2 x° = 1 \Rightarrow x = 0, 180, 360$
hence the least period of $\cos^2 x°$ is 180°.

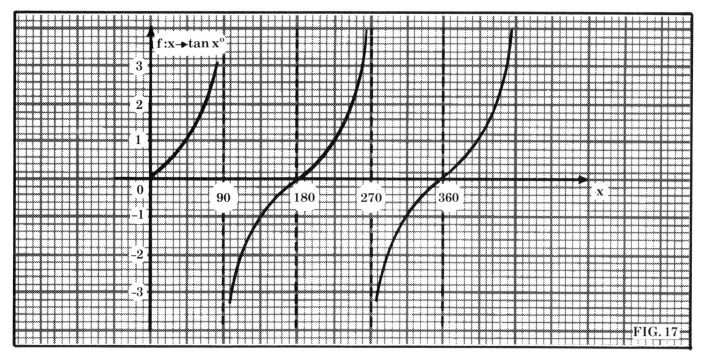

FIG. 17

ASSIGNMENT 1.4

1. In the interval $0 \leqq x \leqq 360$, write down (using the graph where necessary),
   (i) The maximum and minimum values of $\sin x°$ and the value of x at which each occurs.
   (ii) The maximum and minimum values of $\cos x°$ and the value of x at which each occurs.
   (iii) The maximum and minimum values of $\tan x°$ and the value of x at which each occurs.

2. Use the graphs of the sine, cosine and tangent functions to solve the following equations, for the interval $0 \leqq x \leqq 360$,
   (i) $\sin x° = 0.60$ (ii) $\cos x° = 0.60$
   (iii) $\tan x° = -1.20$

3. Draw the graph of the function $f: x \to 2 \sin x° + \cos x°$ for $0 \leqq x \leqq 180$ and use your graph to solve the equation $2 \sin x° + \cos x° = 1.86$.
   Write down from your graph the maximum turning value of the function and the value of x at which it occurs.

4. Draw on separate diagrams the graphs of the following for the given intervals,
   (i) $f: x \to \sin 2x°$, for $0 \leqq x \leqq 90$
   (ii) $f: t \to 2 \cos t°$, for $0 \leqq t \leqq 360$
   (iii) $f: x \to \frac{1}{2} \tan x°$, for $0 \leqq x \leqq 270$

5. Copy and complete the table,

x	0	30	45	60	90	120	135	150	180
$\sin x°$									
$\cos x°$									
$\sin x° + \cos x°$									
$\sin x° - \cos x°$									

Draw (not on the same diagram) the graphs of the functions, $f: x \to \sin x° + \cos x°$ and $f: x \to \sin x° - \cos x°$ for the interval $0 \leq x \leq 180$.

From your graphs write down the solution sets of the equations,

(i) $\sin x° + \cos x° = 0$

(ii) $\sin x° - \cos x° = 0 \cdot 2$

(iii) $\sin x° - \cos x° = 1$

6. The graph shown in figure 18 (opposite) is that of the function $f: t \to a \sin t° - b \cos t°$ where a and b are constants, for values of t from $0°$ to $180°$.

  (i) State the maximum value of the function and the value of t at which it occurs.

  (ii) State the range of t for which $1 \leq a \sin t° - b \cos t° \leq 2$

  (iii) Find the values of a and b

7. The graph of the function $f: x \to p \sin kx°$ where p and k are constants, is shown in figure 19 (opposite) for values of x in the interval $0 \leq x \leq 270$.

  (i) Find from the graph the solution set of the equation $p \sin x° = 0$.

  (ii) Find the values of p and k.

  (iii) For what range of values of x is $2 \leq p \sin kx° \leq 3$.

  (iv) For what range of values of x is $-2 \leq p \sin kx° \leq 0$.

8. Find the least period of,

  (i) $\sin x°, \sin 2x°, \sin^2 x°$.

  (ii) $\tan x°, \tan 2x°, \tan^2 x°$.

9. Find the least period of,

  (i) $2 \cos x°$ (ii) $3 \sin 2x°$ (iii) $-\tan x°$

  (iv) $4 \tan^2 x°$

10. Find the least period of,

  (i) $\cos 3x°$ (ii) $\sin \tfrac{1}{2}x°$ (iii) $\tan \tfrac{1}{2}x°$

  (iv) $\sin 4x°$

POLAR COORDINATES

A point in the x-y plane may be fixed by using,

(i) Cartesian coordinates.

(ii) Polar coordinates.

In **cartesian coordinates**, to fix the point $P(x, y)$ we move a distance x units along the x-axis to the point M and then a distance y units parallel to the y-axis, thus, P is said to have cartesian coordinates (x, y).

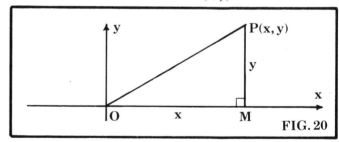

FIG. 20

In **polar coordinates**, to fix the point $P(r, \theta°)$ the line OX is rotated, usually anticlockwise about O through an angle $\theta°$ into a position ON and a length OP equal to r units is measured along ON, thus, P is said to have polar coordinates $(r, \theta°)$.

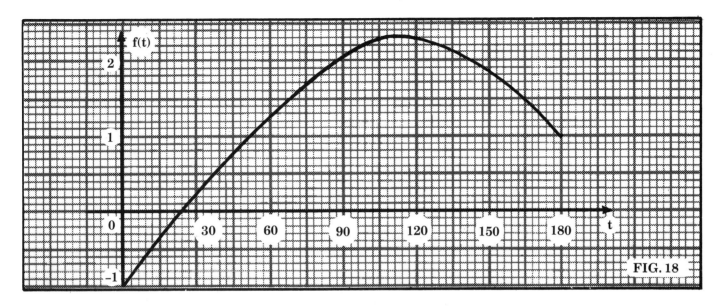

FIG. 18

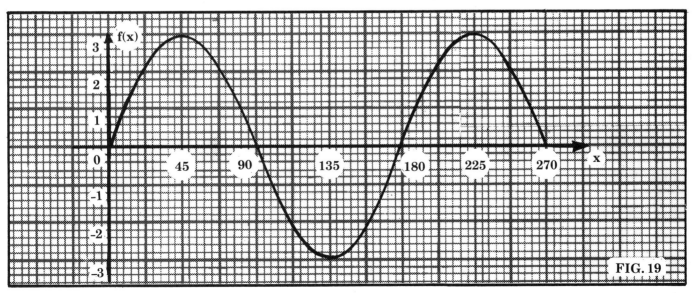

FIG. 19

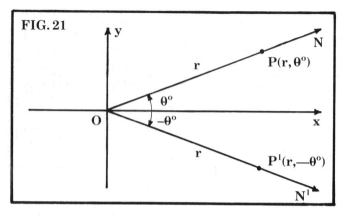

FIG. 21

Note:

(i) OX may be rotated clockwise into a position ON′ (as in figure 21). The angle is then negative. The point P′ will have coordinates $(r, -\theta°)$.

(ii) r is positive if measured along $\overrightarrow{ON}$ and negative if measured from O in the direction $\overrightarrow{NO}$. Thus P″ may have polar coordinates $(-r, \theta°)$ (as in figure 22).

Hence both r and θ may be positive or negative.

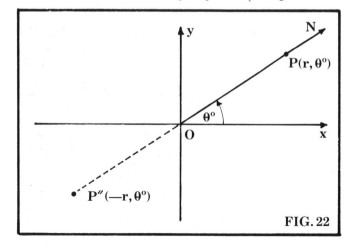

FIG. 22

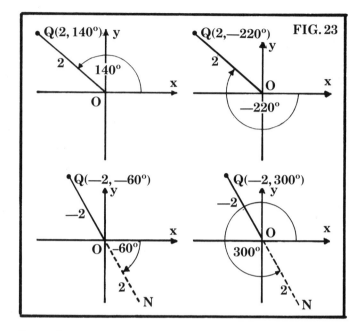

Example 1

The polar coordinates $(2, 140°)$, $(2, -220°)$, $(-2, -60°)$ and $(-2, 300°)$ fix the same point Q, as in figure 23.

In practice it is best to keep both r and θ positive.

To change from Cartesian coordinates to Polar coordinates and vice-versa.

Let the point P be fixed by the cartesian coordinates (x, y) and the polar coordinates $(r, \theta°)$.

From figure 24,

$$\cos \theta° = \frac{x}{r} \Rightarrow x = r \cos \theta°$$

and

$$\sin \theta° = \frac{y}{r} \Rightarrow y = r \sin \theta°$$

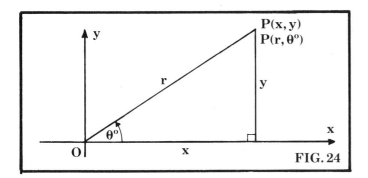

FIG. 24

Hence $x = r\cos\theta$ and $y = r\sin\theta$ are the relations between the cartesian coordinates (x, y) and the polar coordinates $(r, \theta°)$.

To obtain r:
$$x^2 + y^2 = r^2\cos^2\theta° + r^2\sin^2\theta°$$
$$= r^2(\cos^2\theta° + \sin^2\theta°)$$
$$= r^2$$

hence
$$r^2 = x^2 + y^2$$

To obtain θ:
$$\frac{r\sin\theta°}{r\cos\theta°} = \frac{y}{x} \Rightarrow \tan\theta° = \frac{y}{x}$$

Example 2

Find the polar coordinates of the point with cartesian coordinates,

(i) (4, 3)   (ii) $(-\sqrt{3}, -1)$

(i) Here
$$x = r\cos\theta° \Rightarrow 4 = r\cos\theta°$$
$$y = r\sin\theta° \Rightarrow 3 = r\sin\theta°$$
$$\Rightarrow 4^2 + 3^2 = r^2(\cos^2\theta° + \sin^2\theta°)$$
$$\Rightarrow r^2 = 25$$
$$\Rightarrow r = 5 \quad \text{(take the positive value)}$$

and
$$\tan\theta° = \frac{y}{x} = \tfrac{3}{4} = 0.75 \Rightarrow \theta = 36.9$$

i.e. the point (3, 4) has polar coordinates (5, 36.9°).

(ii) Here
$$\left.\begin{array}{l} x = r\cos\theta° = -\sqrt{3} \\ y = r\sin\theta° = -1 \end{array}\right\} \Rightarrow r^2 = 3 + 1$$
$$\Rightarrow r = 2$$
(take the positive value)

and
$$\tan\theta° = \frac{y}{x} = \frac{-1}{-\sqrt{3}}$$

with θ in the 3rd quadrant
$$\Rightarrow \theta = 180° + 30° = 210°$$

i.e. the point $(-\sqrt{3}, -1)$ has polar coordinates $(2, 210°)$.

Example 3

Find the cartesian coordinates of the point with polar coordinates (4, 150°).

Here
$$x = r\cos\theta° = 4\cos 150° = 4 \times \left(-\frac{\sqrt{3}}{2}\right) = -2\sqrt{3}$$

and
$$y = r\sin\theta° = 4\sin 150° = 4 \times \tfrac{1}{2} = 2$$

i.e. the point (4, 150°) has cartesian coordinates $(-2\sqrt{3}, 2)$.

ASSIGNMENT 1.5

1. Plot the points with polar coordinates:
   (i) $(2, 60°)$ (ii) $(1, 155°)$ (iii) $(-3, -150°)$
   (iv) $(-2, 45°)$ (v) $(3, -30°)$ (vi) $(2, 180°)$
   (vii) $(5, 0°)$.

2. Write down three other sets of polar coordinates which will fix each of the following points:
   (i) $(4, 30°)$ (ii) $(-2, 120°)$ (iii) $(-1, -50°)$
   (iv) $(3, -60°)$ (v) $(-3, -240°)$ (vi) $(3, -300°)$.

3. Write down, with r and θ positive, another set of polar coordinates which will fix the points:
   (i) $(3, 390°)$ (ii) $(2, 510°)$ (iii) $(-1, 400°)$
   (iv) $(2, -410°)$ (v) $(-2, -370°)$.

4. Find the polar coordinates of the points whose cartesian coordinates are:
   (i) $(0, 2)$ (ii) $(3, 0)$ (iii) $(\sqrt{3}, 1)$
   (iv) $(1, -\sqrt{3})$ (v) $(\sqrt{2}, -\sqrt{2})$ (vi) $(-1, -1)$
   (vii) $(5, 12)$ (viii) $(-6, 8)$ (ix) $(6, -3)$.

5. Find the cartesian coordinates of the points whose polar coordinates are:
   (i) $(4, 90°)$ (ii) $(2, 180°)$ (iii) $(\sqrt{2}, 135°)$
   (iv) $(\sqrt{3}, 330°)$ (v) $(\sqrt{3}, -120°)$ (vi) $(2, -45°)$
   (vii) $(3, 0°)$ (viii) $(8, 270°)$ (ix) $(4, 200°)$.

1.7 THE SINE RULE

In every triangle ABC,

$$\frac{a}{\sin A} = \frac{b}{\sin B} = \frac{c}{\sin C}$$

Proof: Draw a set of rectangular axes AX and AY, with A as origin and with AC lying along AX.

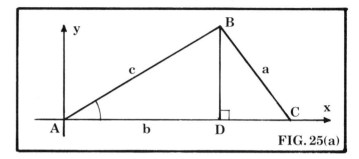
FIG. 25(a)

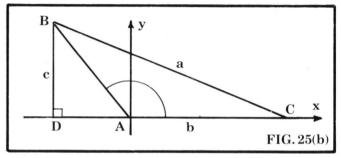
FIG. 25(b)

In figure 25(a) angle A is acute, in figure 25(b) angle A is obtuse. Draw the altitude BD from B to AC.
In both figures,

$$\sin C = \frac{BD}{BC} = \frac{BD}{a} \Rightarrow BD = a \sin C$$

In figure 25(a)

$$\sin A = \frac{BD}{AB} = \frac{BD}{c} \Rightarrow BD = c \sin A$$

In figure 25(b)

$$\sin A = \sin BAD = \frac{BD}{c} \Rightarrow BD = c \sin A$$

Hence in both figures

$$a \cdot \sin C = c \cdot \sin A$$

$$\Rightarrow \frac{a}{\sin A} = \frac{c}{\sin C}$$

By placing AB along AX, we get

$$\frac{a}{\sin A} = \frac{b}{\sin B}$$

Hence

$$\frac{a}{\sin A} = \frac{b}{\sin B} = \frac{c}{\sin C}$$

Note:

(i) The sine rule is used if we are given
  (a) two angles and one side or
  (b) two sides and an angle opposite one of the given sides.

(ii) The following form of the sine rule is more convenient when an angle is to be found

$$\frac{\sin A}{a} = \frac{\sin B}{b} = \frac{\sin C}{c}$$

(iii) If one of the angles is obtuse, e.g. 135°, use the fact that $\sin 135° = \sin(180 - 135)° = \sin 45°$.

Example 1

In triangle ABC, $\hat{A} = 38°$, $\hat{B} = 79°$ and $a = 15 \cdot 4$. Calculate the length of the sides b and c.

$$\hat{A} + \hat{B} + \hat{C} = 180° \Rightarrow \hat{C} = 180° - (\hat{A} + \hat{B})$$
$$= 180° - 117° = 63°$$

By the sine rule

$$\frac{a}{\sin A} = \frac{b}{\sin B} = \frac{c}{\sin C} \Rightarrow \frac{15 \cdot 4}{\sin 38°} = \frac{b}{\sin 79°} = \frac{c}{\sin 63°}$$

the first and second ratios $\Rightarrow b = \dfrac{15 \cdot 4 \sin 79°}{\sin 38°}$

the first and third ratios $\Rightarrow c = \dfrac{15 \cdot 4 \sin 63°}{\sin 38°}$

No.	log
15·4	1·188
sin 79°	$\overline{1}$·992
add	1·180
sin 38°	$\overline{1}$·789
24·6	1·391

No.	log
15·4	1·188
sin 63°	$\overline{1}$·950
add	1·138
sin 38°	$\overline{1}$·789
22·3	1·349

i.e. $b = 24 \cdot 6$ and $c = 22 \cdot 3$.

Example 2

A lighthouse L and two buoys B_1 and B_2 mark a dangerous stretch of water. L is north-east of B_1 and B_1 is west of B_2. The bearing of B_2 from L is 160°. The distance between B_1 and B_2 is 1620 metres. Calculate the distances LB_1 and LB_2 to three significant figures.

In questions involving bearings it is essential to start by drawing a set of rectangular axes to represent the points of the compass N, S, E, W as in figure 26 (over page). Then take one of the given points as origin, in this case B_1, and fix the second point L.

If a particular point B_2 is given as a circular bearing from another point L, then **draw** in the north-line at L namely LN. *Circular bearings are always measured clockwise from the North line.*

Lastly mark in the other angles of your triangle.

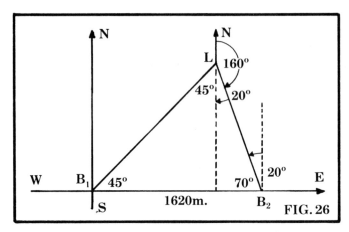

FIG. 26

By the sine rule

$$\frac{B_1B_2}{\sin 65°} = \frac{LB_1}{\sin B_2} = \frac{LB_2}{\sin B_1}$$

$$\Rightarrow \frac{1620}{\sin 65°} = \frac{LB_1}{\sin 70°} = \frac{LB_2}{\sin 45°}$$

the first and second ratios $\Rightarrow LB_1 = \dfrac{1620 \sin 70°}{\sin 65°}$

the first and third ratios $\Rightarrow LB_2 = \dfrac{1620 \sin 45°}{\sin 65°}$

No.	log	No.	log
1620	3·210	1620	3·210
sin 70°	$\bar{1}$·973	sin 45°	$\bar{1}$·849
add	3·183	add	3·059
sin 65°	$\bar{1}$·957	sin 65°	$\bar{1}$·957
1680	3·226	1260	3·102

$\therefore LB_1 = 1680\,\text{m} \qquad LB_2 = 1260\,\text{m}$

THE COSINE RULE

In every triangle ABC, $a^2 = b^2 + c^2 - 2bc \cos A$.

Proof: Draw a set of rectangular axes AX and AY with A as origin and with AC lying along AX.

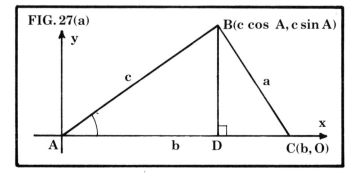

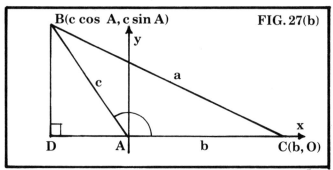

In figure 27(a) the angle A is acute, in figure 27(b) angle A is obtuse. Draw the altitude BD from B to AC.

With the usual notation $AC = b$, $\therefore$ C has coordinates $(b, 0)$.

In figure 27(a)

$AD = AB \cos A = c \cos A$ and $DB = AB \sin A = c \sin A$

$\therefore$ B has coordinates $(c \cos A, c \sin A)$

In figure 27(b)
$$\cos A = \frac{AD}{AB} \Rightarrow AD = c \cos A$$
and
$$\sin A = \frac{DB}{AB} \Rightarrow DB = c \sin A$$

∴ B has coordinates $(c \cos A, c \sin A)$

If B is the point (x_2, y_2) and C the point (x_1, y_1), then by the distance formula, $BC^2 = (x_2 - x_1)^2 + (y_2 - y_1)^2$. For the points $B(c \cos A, c \sin A)$, and $C(b, 0)$
$$BC^2 = (c \cos A - b)^2 + (c \sin A - 0)^2 \quad \text{and} \quad BC = a$$
hence,
$$a^2 = c^2 \cos^2 A - 2bc \cos A + b^2 + c^2 \sin^2 A$$
$$= b^2 + c^2 \sin^2 A + c^2 \cos^2 A - 2bc \cos A$$
$$= b^2 + c^2(\sin^2 A + \cos^2 A) - 2bc \cos A$$
$$= b^2 + c^2 - 2bc \cos A \quad (\text{since } \sin^2 A + \cos^2 A = 1)$$
$$\therefore a^2 = b^2 + c^2 - 2bc \cos A$$
and
$$\left.\begin{array}{l} b^2 = c^2 + a^2 - 2ca \cos B \\ c^2 = a^2 + b^2 - 2ab \cos C \end{array}\right\} \text{these are other forms.}$$

Note:
(i) The cosine rule is used to find a side or an angle when,
  (a) two sides and the included angle are given.
  (b) three sides are given.
(ii) It is sometimes necessary to rewrite the cosine formulae in another form, namely,
$$\cos A = \frac{b^2 + c^2 - a^2}{2bc}$$
$$\cos B = \frac{c^2 + a^2 - b^2}{2ca}$$
$$\cos C = \frac{a^2 + b^2 - c^2}{2ab}$$
when the three sides are given and it is required to find the size of an angle.

(iii) If one of the angles is obtuse, e.g. 135°, use the fact that $\cos 135° = -\cos(180 - 135)° = -\cos 45°$.

Example 3

Calculate the side c of triangle ABC, with $a = 12$, $b = 13$, $\hat{C} = 10°$.

Use
$$c^2 = a^2 + b^2 - 2ab \cos C$$
$$= 144 + 169 - 2.12.13.\cos 10°$$
$$= 313 - 312 \cos 10°$$
$$= 313 - 307$$
$$= 6$$
hence
$$c = \sqrt{6} = 2·45$$

Example 4

In triangle ABC, $a = 10$, $b = 15$, $c = 8$, calculate the size of angle B.

Use
$$\cos B = \frac{c^2 + a^2 - b^2}{2ca} = \frac{64 + 100 - 225}{2.10.8}$$
$$= \frac{164 - 225}{160} = -\frac{61}{160}$$
and
$$\cos B = -\frac{61}{160} \Rightarrow B = 180° - 67·6°$$
$$= 112·4°$$

THE AREA OF TRIANGLE ABC = $\frac{1}{2}ab \sin c$

Let the area of triangle ABC be denoted by $\triangle$.

Proof: In figure 28(a) angle C is acute, in figure 28(b) angle C is obtuse. Let altitude BD = h.

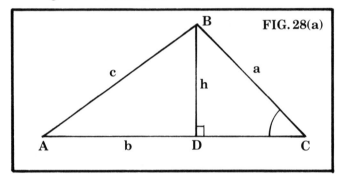

FIG. 28(a)

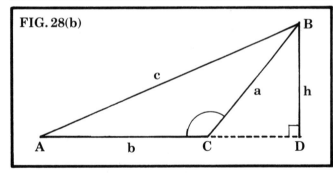

FIG. 28(b)

$\triangle = \frac{1}{2}$ base × height = $\frac{1}{2}$AC × h = $\frac{1}{2}b \times h$

but $h = a \sin C$ in figure 28(a) and $h = a \sin$ BCD = $a \sin C$ in figure 28(b).

$$\therefore \triangle = \tfrac{1}{2}ab \sin C$$

Other forms are

$$\triangle = \tfrac{1}{2}bc \sin A$$
$$= \tfrac{1}{2}ca \sin B$$

ASSIGNMENT 1.6

1. Triangle ABC has A = 45·2°, B = 58·1° and c = 7·41. Calculate the length of side b.

2. Find the size of the greatest angle of the triangle with sides of length 5·1 cm, 4·7 cm and 7·2 cm.

3. A triangle has sides 12 units, 13 units and 18 units. Find the size of the smallest angle and the area of the triangle.

4. Triangle ABC has a = 4·8 cm, b = 5·2 cm and has area 7·58 cm². Calculate the possible sizes of angle C.

5. State the area of $\triangle$PQR in terms of p, q and R. By using the sine rule, prove that the area of $\triangle$PQR can be written as

$$\frac{p^2 \sin Q \sin R}{2 \sin P}$$

If $\triangle$PQR has area 65·4 cm² and has two angles of 30·6° and 49·7°, calculate the length of the largest side of the triangle.

6. Ship A is 14 km from a port C on a bearing 022°. Ship B is 17 km from port C on a bearing 085°. Calculate the distance between the ships to the nearest km.

7. With reference to rectangular axes OX, OY, OZ in 3 dimensions, A has coordinates (0, 0, 5), B(12, 0, 0) and C(0, 8, 0). Calculate the size of the angle BAC.

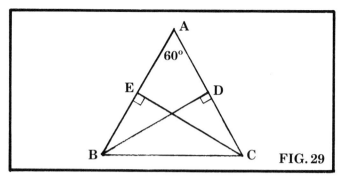

FIG. 29

8. In triangle ABC figure 29, angle A = 60°. BD and CE are altitudes from B and C to AC and AB respectively. Prove with the usual notation, that,

   (i) $AE = b/2$.

   (ii) by applying the cosine rule to triangles ABC and AED, $BC = 2ED$.

   (iii) area of $\triangle ABC = 4 \cdot$ area of $\triangle AED$.

9. The diagram in figure 30 shows a flagpole d units in length standing at the top of a hill. A man at the foot of the hill observes that the angle of elevation of the foot of the pole is α and that the angle of elevation of the top of the pole is β.

   If CD represents the flagpole, angle $BAC = \alpha$ and angle $BAD = \beta$ the angles of elevation from A to the foot and top of the flagpole,

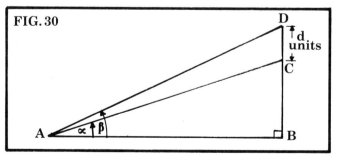

FIG. 30

   (i) show that sin DCA = cos α.

   (ii) show by applying the sine rule to triangle ADC that
   $$AD = \frac{d \cos \alpha}{\sin(\beta - \alpha)}$$
   and find a similar expression for AC.

10. Triangle PQR is an isosceles triangle with $PQ = b = PR$, and angle $QPR = 2\alpha$. Show that $QR = 2b \sin \alpha$.

    If QR is produced its own length to S, prove by applying the cosine rule to triangle PQS that,
    $$PS = b\sqrt{(1 + 8 \sin^2 \alpha)}$$

11. Triangle ABC is isosceles with $BC = a$ units and $AB = AC = na$ units. Show that
    $$\cos BAC = \frac{2n^2 - 1}{2n^2}$$
    and that the area of triangle $ABC = \tfrac{1}{4}a^2\sqrt{(4n^2 - 1)}$.

12. From two stations P and Q on the sea-shore 4 km apart a ship S is observed at sea. The bearing of S from P is 044° and from Q is 323° (measured clockwise from the north). Calculate in metres to three significant figures the distance of the ship from both stations.

13. An aircraft leaves an airfield R and flies 138 km on a course of 105° to a point S. It then changes course and flies 96 km on a course of 190° to a point T. Calculate the distance of T from R and the bearing of T from R.

14. P is the point with polar coordinates $(r_1, \theta_1°)$ and Q the point with polar coordinates $(r_2, \theta_2°)$. Show that the distance PQ is given by,
    $$PQ^2 = r_1^2 + r_2^2 - 2r_1 r_2 \cos(\theta_1 - \theta_2)°.$$

15. Calculate the length of the line segments joining the following pairs of points whose coordinates are given in polar form.

   (i) $(2, 60°)$ and $(3, 120°)$

   (ii) $(3, 212°)$ and $(2, 122°)$

   (iii) $(3, 30°)$ and $(3, 150°)$

   (iv) $(4, 252°)$ and $(1, 72°)$.

16. M and N are variable points in the plane with polar coordinates $M(2, \theta_1°)$, $30 \leq \theta_1 \leq 60$ and $N(3, \theta_2°)$, $120 \leq \theta_2 \leq 200$. Find the greatest and least distances between the points M and N.

17. Show that the triangle with vertices the points $(0, 0)$, $(4, 52°)$ and $(4, 112°)$ is equilateral.

ASSIGNMENT 1.7

Objective items testing Sections 1.1–1.7
Instructions for answering these items are given on page 16.

1. If $\sin x° < 0$ and $\cos x° > 0$ then x lies in the interval,
   A. $180 < x < 270$
   B. $180 < x < 360$
   C. $270 < x < 360$
   D. $0 < x < 180$
   E. $90 < x < 180$

2. If $\cos(-x°) = -\sin 210°$ for $0 \leqq x \leqq 180$ then x has the value
   A. 30
   B. 60
   C. 90
   D. 120
   E. 150

3. If $\tan \theta = -\frac{3}{4}$ and $0 \leqq \theta \leqq 180$, then the exact value of $\dfrac{\sin \theta - \cos \theta}{\sin \theta + \cos \theta}$ is
   A. -7
   B. $-\frac{1}{7}$
   C. $\frac{1}{7}$
   D. 7
   E. None of these values.

4. When simplified $\dfrac{\cos P}{1 - \sin P} - \dfrac{\cos P}{1 + \sin P}$ is equal to
   A. $\dfrac{2}{\cos P}$
   B. $2 \tan P$
   C. $\dfrac{2 \cos P}{\sin^2 P}$
   D. $\dfrac{2 \cos P}{1 - \sin P}$
   E. $\dfrac{2 \sin P}{1 - \sin^2 P}$

5. The point P has polar coordinates $(5, 20°)$. Its image under reflection in the x-axis is
   A. $(5, 160°)$
   B. $(5, -20°)$
   C. $(-5, 20°)$
   D. $(-5, -20°)$
   E. $(-5, -160°)$

6. The value of x for which $\cos(x+30)°$ takes its maximum value in the range $0 \leqq x \leqq 360$ is
   A. 30
   B. 150
   C. 240
   D. 330
   E. None of these.

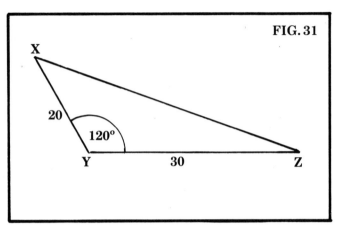

FIG. 31

10. Triangle PQR has PQ = 4·5 cm, QR = 8 cm and area = 9 cm². Possible value(s) of angle Q is/are,
    (1) 30°
    (2) 60°
    (3) 150°

11. (1) PQR is any triangle
    (2) $\sin P = \sin(Q+R)$

12. (1) In a triangle PQR, $\sin Q = \sin R$
    (2) $\hat{Q} = \hat{R} = 45°$

7. In triangle XYZ (above), XZ is equal to
   A. $10\sqrt{13}$
   B. 40
   C. $10\sqrt{3}+20\sqrt{2}$
   D. $10\sqrt{19}$
   E. $10\sqrt{7}$

8. Eliminating θ from the equations $y = \cos\theta + \sin\theta$ and $x = 2\sin\theta$ gives
   A. $(2y-x)^2 + x^2 = 1$
   B. $(y-x)^2 = 2$
   C. $y^2 + 2xy + x^2 = 1$
   D. $2y^2 - 2xy + x^2 = 2$
   E. $(x-2y)^2 + x^2 = 2$

9. Which of the following functions have a least period of 2π?
   (1) $\sin x$
   (2) $\sin 2x$
   (3) $\sin^2 x$

UNIT 2: CIRCULAR MEASURE

INTRODUCTION 2.1

Let an arc AB subtend an angle of $x°$ at the centre O of a circle of radius r, then by proportion,

$$\frac{\text{length of the arc AB}}{\text{length of the circumference}} = \frac{x°}{360°}$$

i.e.

$$\frac{\text{arc AB}}{2\pi r} = \frac{x°}{360°} \Rightarrow \text{arc AB} = \frac{\pi r x°}{180°}$$

$$\Rightarrow \frac{\text{arc AB}}{r} = \frac{\pi x°}{180°}$$

Let the arcs AB, A_1B_1 and A_2B_2 of the three circles in figure 33 with radii r, r_1 and r_2 respectively subtend an angle of $x°$ at the centres O, O_1 and O_2

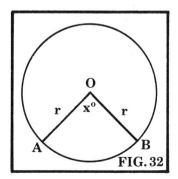

FIG. 32

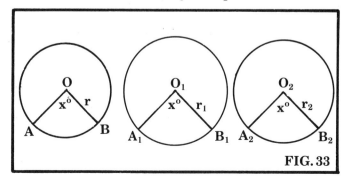

FIG. 33

then

$$\frac{\text{arc AB}}{r} = \frac{\pi x°}{180°}$$

$$\frac{\text{arc } A_1 B_1}{r_1} = \frac{\pi x°}{180°}$$

$$\frac{\text{arc } A_2 B_2}{r_2} = \frac{\pi x°}{180°}$$

hence

$$\frac{\text{arc AB}}{r} = \frac{\text{arc } A_1 B_1}{r_1} = \frac{\text{arc } A_2 B_2}{r_2}$$

and this value will be the same for all circles, provided the angle subtended by the arcs is the same. Hence for this reason,

$\dfrac{\text{arc AB}}{r}$ is called the **radian measure** of the angle AOB

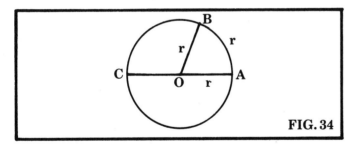

FIG. 34

If arc AB = r (the radius of the circle) as in figure 34 then

$$\frac{\text{arc AB}}{r} = \frac{r}{r} = 1$$

and we define $\angle$ **AOB = 1 Radian**.

Hence the definition:

One Radian is the angle subtended at the centre of a circle by an arc equal in length to the radius of the circle.

Also, since the length of an arc of a circle is proportional to the size of the angle it subtends at the centre, we get from figure 34 with AOC as a diameter,

$$\frac{\text{arc AB}}{\angle \text{AOB}} = \frac{\text{arc AC}}{\angle \text{AOC}}$$

but

arc AB = radius = r

arc AC = length of the circumference of a semicircle = πr

$\angle$ AOB = 1 Radian
$\angle$ AOC = 180°

$\therefore$ we get $\dfrac{r}{1} = \dfrac{\pi r}{180°}$

and hence

180° = π Radians

Note:

$$90° = \frac{\pi}{2} \text{ radians}$$

$$60° = \frac{\pi}{3} \text{ radians}$$

$$30° = \frac{\pi}{6} \text{ radians}$$

$$42° = \frac{42\pi}{180} \text{ radians}$$

$$\therefore x° = \frac{\pi x}{180} \text{ radians.}$$

Also

$$\pi \text{ radians} = 180° \Rightarrow 1 \text{ radian} = \left(\frac{180}{\pi}\right)° \doteqdot 57\cdot 3°$$

TRIGONOMETRICAL FUNCTIONS IN RADIANS

Since

$$x° = \frac{\pi x}{180} \text{ radians} \Rightarrow x \text{ radians} = \left(\frac{180x}{\pi}\right)°$$

then

$$\sin x° = \sin\left(\frac{\pi x}{180}\right)$$

and

$$\sin x = \sin\left(\frac{180x}{\pi}\right)°$$

Therefore in trigonometry $\sin x°$ is the sine of the angle whose size is x **degrees** where x is a real number and $\sin x$ is the sine of the real number x which is the sine of the angle whose **radian** measure is x.

$$\therefore \sin x \neq \sin x°$$

Similarly for $\cos x$ and $\tan x$.

Example 1
(i) Express in radians 54°
(ii) Express in degrees 1·8 radians.

(i) $\qquad 180° = \pi$ radians

$$\Rightarrow 54° = \frac{\pi}{180} \times 54 \text{ radians}$$

$$= \frac{\pi \times 6}{20}$$

$$= \frac{3 \cdot 14 \times 3}{10}$$

$$= 0 \cdot 942 \text{ radians}$$

(ii) $\qquad \pi$ radians = 180°

$$\Rightarrow 1 \cdot 8 \text{ radians} = \left(\frac{180}{\pi} \times 1 \cdot 8\right)°$$

$$= \left(\frac{180}{\pi} \times \frac{18}{10}\right)°$$

$$= \left(\frac{18 \times 18}{3 \cdot 14}\right)°$$

$$= 103 \cdot 18°$$

OR 1 radian $\doteqdot 57 \cdot 3°$

$$\Rightarrow 1 \cdot 8 \text{ radians} \doteqdot (57 \cdot 3 \times 1 \cdot 8)°$$

$$= 103 \cdot 14°$$

Example 2
Find the solution sets of the equations,

(i) $\qquad \cos x = \tfrac{1}{2}, \qquad 0 \leqq x < 2\pi$

(ii) $\qquad \tan x = -\dfrac{1}{\sqrt{3}}, \qquad 0 \leqq x < 2\pi$

(i) *Note*: $\cos x° = \tfrac{1}{2} \Rightarrow x = 60$ or 300, so for $0 \leqq x < 360$ the solution set is $\{60, 300\}$.

$$\therefore \cos x = \frac{1}{2} \Rightarrow x = \frac{\pi}{3} \text{ or } \frac{5\pi}{3}, \text{ so for } 0 \leqq x < 2\pi \text{ the solution set is } \left\{\frac{\pi}{3}, \frac{5\pi}{3}\right\}.$$

(ii) *Note*: $\tan x° = -\dfrac{1}{\sqrt{3}} \Rightarrow x = 150$ or 330, so for $0 \leqq x < 360$ the solution set is $\{150, 330\}$.

$$\therefore \tan x = -\frac{1}{\sqrt{3}} \Rightarrow x = \frac{5\pi}{3} \text{ or } \frac{11\pi}{6}, \text{ so for } 0 \leqq x < 2\pi \text{ the solution set is } \left\{\frac{5\pi}{3}, \frac{11\pi}{6}\right\}.$$

Example 3

A chord AB of a circle of radius r subtends an angle of θ radians at the centre.

(i) Find in terms of r and θ the area of the smaller segment.

(ii) If this area is $\frac{1}{4}$ the area of the whole circle prove that $\theta - \sin\theta = \frac{\pi}{2}$.

(i) The smaller segment ACB is denoted by the shaded area in figure 35.

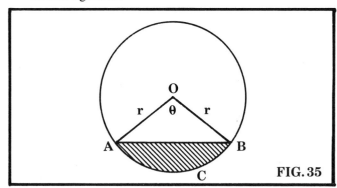

FIG. 35

$\therefore$ Area of segment ACB =
   area of the sector AOBC − area of triangle AOB

Since the area of a sector is proportional to the angle subtended at the centre, then

$$\frac{\text{area of AOBC}}{\text{area of circle}} = \frac{\theta}{2\pi}$$

$\therefore$ area of AOBC $= \dfrac{\theta \times \pi r^2}{2\pi} = \tfrac{1}{2}r^2\theta$

and area of triangle AOB $= \tfrac{1}{2}r^2 \sin\theta$

$\therefore$ Area of the smaller segment $= \tfrac{1}{2}r^2\theta - \tfrac{1}{2}r^2 \sin\theta$
$\qquad\qquad\qquad\qquad\qquad = \tfrac{1}{2}r^2(\theta - \sin\theta)$

(ii) Here

$$\tfrac{1}{2}r^2(\theta - \sin\theta) = \tfrac{1}{4}\pi r^2$$

$$\Leftrightarrow \tfrac{1}{2}(\theta - \sin\theta) = \frac{\pi}{4}$$

$$\Leftrightarrow \quad \theta - \sin\theta = \frac{\pi}{2}$$

ASSIGNMENT 2.1

$\left[\text{Note: } 180° = \pi \text{ radians}, \ x° = \dfrac{\pi x}{180} \text{ radians}, \ x \text{ radians} = \left(\dfrac{180x}{\pi}\right)°\right].$

1. Copy and complete the following tables:

angle in degrees	0	30	60	180	120	150
angle in radians			$\pi/3$			

angle in degrees					
angle in radians	$\pi/4$	$\pi/2$	$3\pi/4$	π	2π

2. Express in terms of π the radian measure of,
   40°, 81°, 108°, 225°, 540°.

3. Express in degrees the following radian measures,
   $\dfrac{\pi}{8}$, 2, $\dfrac{5\pi}{9}$, 1·5, 3·4.

4. Without using tables find the values of,

   (i) $\sin\dfrac{\pi}{4}$ (ii) $\cos\dfrac{2\pi}{3}$ (iii) $\tan\pi$

(iv) $\tan \dfrac{5\pi}{6}$ (v) $\sin\left(-\dfrac{\pi}{3}\right)$ (vi) $\cos \dfrac{7\pi}{6}$

5. Give the radian measure of each of the following (take $\pi = 3\cdot 14$)

   $43°$, $72°$, $110°$, $400°$.

6. Find in degrees the size of the angles whose radian measure is,

   $0\cdot 45$, $0\cdot 8$, $1\cdot 4$, $2\cdot 51$.

7. Simplify,
   (i) $\cos(\tfrac{1}{2}\pi - \theta)$ (ii) $\tan(\pi - \theta)$ (iii) $\sin(2\pi - \theta)$
   (iv) $\sin(\pi + \theta)$ (v) $\cos(2\pi + \theta)$.

8. Give the solution set of the following equations for the interval $0 \leq x \leq 2\pi$. (Answers in radian measure).
   (i) $\sin x = \dfrac{\sqrt{3}}{2}$ (ii) $\cos x = 1$ (iii) $\tan x = \tan 45°$
   (iv) $\sin x = 0$ (v) $\tan x = -\sqrt{3}$

9. Give in radian measure the solution set of each of the equations for the interval $0 \leq \theta \leq 2\pi$.
   (i) $\sin(\tfrac{1}{2}\pi - \theta) = 1$ (ii) $\cos(\pi + \theta) = -\tfrac{1}{2}$
   (iii) $\tan(2\pi - \theta) = 0$.

10. An arc AB of a circle of radius r subtends an angle $x°$ at the centre. State in terms of r and x the formula for the length of the arc AB, and hence prove that if the angle subtended by the arc AB at the centre has radian measure θ, then the length of the arc AB = $r\theta$ (see figure 36).

11. An arc AB of a circle of radius r subtends an angle of $x°$ at the centre O.

    Prove (using proportion) that the area of the sector

    $$\text{AOB} = \dfrac{\pi r^2 x}{360}$$

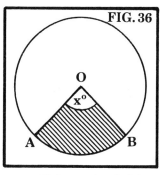

FIG. 36

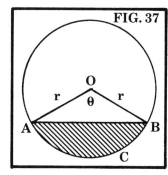

FIG. 37

If the angle of $x°$ is now written as θ radians, show that the area of the sector AOB can now be written as $\tfrac{1}{2}r^2\theta$.

12. In a circle of radius 5 cm an arc AB is 8 cm in length. Find the angle subtended by the arc at the centre,
    (1) in radians   (ii) in degrees.
    (take $\pi = 3\cdot 14$ and use the formula obtained in question 10.)

13. A chord AB of a circle centre O and of radius r, subtends an angle of θ radians at the centre (see figure 37).
    Prove that the area of the smaller segment ACB of the circle is given by $\tfrac{1}{2}r^2(\theta - \sin \theta)$.
    (Use the fact that area of segment ACB = area of sector AOBC − area of $\triangle$AOB).
    If AB divides the circle into two segments whose areas are in the ratio 1:2, show that

    $$\theta - \sin \theta = \dfrac{2\pi}{3}$$

14. Two circles of equal radius r cm intersect and have their centres a distance r cm apart.
    Show that the area common to both circles is

    $$r^2\left(\dfrac{2\pi}{3} - \dfrac{\sqrt{3}}{2}\right).$$

Show also that the common chord divides each circle into two parts whose areas are in the ratio
$$\frac{4\pi - 3\sqrt{3}}{8\pi + 3\sqrt{3}}$$

ASSIGNMENT 2.2

Objective items testing Sections 2.1 and 2.2.
Instructions for answering these items are given on page 16.

1. The value of $\sin \frac{5\pi}{6}$ is

   A. $-\frac{\sqrt{3}}{2}$

   B. $-\frac{1}{2}$

   C. $\frac{1}{\sqrt{3}}$

   D. $\frac{1}{2}$

   E. $\frac{\sqrt{3}}{2}$

2. The equation $\cos^2 x = \frac{1}{4}$ for $0 \leq x \leq 2\pi$ has solution set

   A. $\left\{\frac{\pi}{3}, \frac{2\pi}{3}\right\}$

   B. $\left\{\frac{\pi}{3}, \frac{2\pi}{3}, \frac{4\pi}{3}, \frac{5\pi}{3}\right\}$

   C. $\left\{\frac{\pi}{3}, \frac{2\pi}{3}, \pi, \frac{4\pi}{3}\right\}$

   D. $\left\{\frac{\pi}{3}, \frac{5\pi}{3}, \frac{2\pi}{3}\right\}$

   E. $\left\{\frac{4\pi}{3}, \frac{5\pi}{3}\right\}$

3. An arc PQ of a circle subtends an angle of 120° at the centre O. The length of the arc PQ as a fraction of the length of the circumference of the circle is

   A. $\frac{1}{8}$

   B. $\frac{1}{5}$

   C. $\frac{1}{4}$

   D. $\frac{1}{3}$

   E. Not enough information is given.

4. The value of $\dfrac{\sin \frac{5\pi}{6}}{\cos \frac{5\pi}{6}}$ is

   A. $-\sqrt{3}$

   B. $-\frac{1}{\sqrt{3}}$

   C. 1

   D. $\frac{\sqrt{3}}{2}$

   E. $\frac{1}{\sqrt{3}}$

5. If $\cos x = -\frac{\sqrt{3}}{2}$ for $0 \leq x \leq \pi$, x is

   A. $\frac{\pi}{3}$

   B. $\frac{\pi}{6}$

   C. $\frac{2\pi}{3}$

   D. $\frac{5\pi}{6}$

   E. No value exists.

6. If $\cos\left(\dfrac{2\pi}{3} + \theta\right) = -\dfrac{1}{2}$ and $0 < \theta < \pi$, θ is

   A. $\dfrac{\pi}{3}$

   B. $-\dfrac{\pi}{3}$

   C. $-\dfrac{2\pi}{3}$

   D. $\dfrac{2\pi}{3}$

   E. No possible value for θ in the range $0 < \theta < \pi$.

7. The length of an arc PQ of a circle is an eighth of the length of the circumference of the circle with centre O. The size of the angle POQ in radians is

   A. $\dfrac{\pi}{8}$

   B. $\dfrac{\pi}{4}$

   C. π

   D. $\dfrac{\pi}{2}$

   E. It cannot be found without further information.

8. Sector POQ is an eighth of the area of a circle with centre O. The size of the angle POQ in radians is

   A. $\dfrac{\pi}{16}$

   B. $\dfrac{\pi}{8}$

   C. $\dfrac{\pi}{4}$

   D. $\dfrac{\pi}{2}$

   E. Unable to be calculated without further information.

9. Which of the following angles are members of the solution set of the equation $\sin\alpha(2\cos\alpha + 1) = 0$ for $0 \leq \alpha \leq 2\pi$?

   (1) $\dfrac{\pi}{3}$

   (2) $\dfrac{2\pi}{3}$

   (3) $\dfrac{4\pi}{3}$

10. If $\cos\left(\dfrac{180}{\pi}x\right)^\circ = \dfrac{1}{2}$, $0 \leq x \leq 360$, which of the following are values of x?

    (1) $\dfrac{\pi}{3}$

    (2) $\dfrac{2\pi}{3}$

    (3) $\dfrac{4\pi}{3}$

11. (1) $\cos\left(x - \frac{\pi}{2}\right) = \frac{1}{2}$, $0 \leq x < 2\pi$.

   (2) $x = \frac{2\pi}{3}$

12. (1) $\sin x = \cos x$, $0 \leq x < 2\pi$.

   (2) $x = \frac{\pi}{4}$

2.3 THE ROTATION OF THE PLANE THROUGH AN ANGLE θ ABOUT THE ORIGIN O

Consider a rotation of the plane through an angle θ radians about the origin O. By this rotation the point $P(x, 0)$ maps onto $P'(x \cos \theta, x \sin \theta)$.

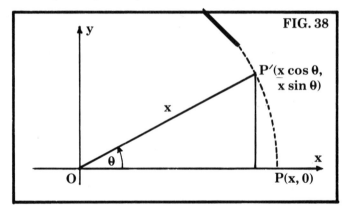

FIG. 38

If R is the matrix associated with this rotation, then

$$R\begin{pmatrix} x \\ 0 \end{pmatrix} = \begin{pmatrix} x \cos \theta \\ x \sin \theta \end{pmatrix}$$

Let

$$R = \begin{pmatrix} a & b \\ c & d \end{pmatrix}$$

then

$$R\begin{pmatrix} x \\ 0 \end{pmatrix} = \begin{pmatrix} a & b \\ c & d \end{pmatrix}\begin{pmatrix} x \\ 0 \end{pmatrix} = \begin{pmatrix} ax \\ cx \end{pmatrix}$$

hence

$$\begin{pmatrix} ax \\ cx \end{pmatrix} = \begin{pmatrix} x \cos \theta \\ x \sin \theta \end{pmatrix} \Rightarrow \begin{cases} a = \cos \theta \text{ and} \\ c = \sin \theta \end{cases}$$

By the same rotation $Q(0, y) \to Q'(-y \sin \theta, y \cos \theta)$.

Hence

$$R\begin{pmatrix} 0 \\ y \end{pmatrix} = \begin{pmatrix} -y \sin \theta \\ y \cos \theta \end{pmatrix}$$

$$\therefore \begin{pmatrix} a & b \\ c & d \end{pmatrix}\begin{pmatrix} 0 \\ y \end{pmatrix} = \begin{pmatrix} -y \sin \theta \\ y \cos \theta \end{pmatrix}$$

$$\Rightarrow \begin{pmatrix} by \\ dy \end{pmatrix} = \begin{pmatrix} -y \sin \theta \\ y \cos \theta \end{pmatrix} \Rightarrow \begin{cases} b = -\sin \theta \text{ and} \\ d = \cos \theta \end{cases}$$

$$\therefore R = \begin{pmatrix} a & b \\ c & d \end{pmatrix} = \begin{pmatrix} \cos \theta & -\sin \theta \\ \sin \theta & \cos \theta \end{pmatrix}$$

and is the matrix associated with the operation of rotation of the plane through an angle θ about the origin.

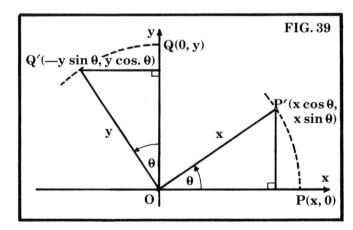

FIG. 39

THE ADDITION FORMULAE—PART (i)
THE FORMULAE FOR cos $(\alpha + \beta)$ AND sin $(\alpha + \beta)$

Let P be any point with coordinates (x, y). Let $P(x, y) \to P'(x', y')$ by a rotation of α radians about O.

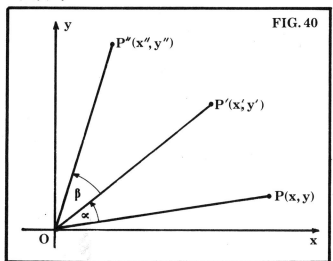

FIG. 40

Therefore, the matrix associated with this rotation is

$$R_1 = \begin{pmatrix} \cos \alpha & -\sin \alpha \\ \sin \alpha & \cos \alpha \end{pmatrix}$$

hence

$$R_1 \begin{pmatrix} x \\ y \end{pmatrix} = \begin{pmatrix} x' \\ y' \end{pmatrix}$$

Let $P'(x', y') \to P''(x'', y'')$ by a rotation of β radians about O. Hence

$$R_2 \begin{pmatrix} x' \\ y' \end{pmatrix} = \begin{pmatrix} x'' \\ y'' \end{pmatrix}$$

where

$$R_2 = \begin{pmatrix} \cos \beta & -\sin \beta \\ \sin \beta & \cos \beta \end{pmatrix}$$

$$\therefore R_2 R_1 \begin{pmatrix} x \\ y \end{pmatrix} = \begin{pmatrix} x'' \\ y'' \end{pmatrix}$$

But $P(x, y) \to P''(x'', y'')$ by a rotation of $\alpha + \beta$ radians about O, hence

$$R \begin{pmatrix} x \\ y \end{pmatrix} = \begin{pmatrix} x'' \\ y'' \end{pmatrix},$$

where

$$R = \begin{pmatrix} \cos(\alpha+\beta) & -\sin(\alpha+\beta) \\ \sin(\alpha+\beta) & \cos(\alpha+\beta) \end{pmatrix}$$

$$\therefore R = R_2 R_1$$

i.e.

$$\begin{pmatrix} \cos(\alpha+\beta) & -\sin(\alpha+\beta) \\ \sin(\alpha+\beta) & \cos(\alpha+\beta) \end{pmatrix}$$
$$= \begin{pmatrix} \cos \beta & -\sin \beta \\ \sin \beta & \cos \beta \end{pmatrix} \begin{pmatrix} \cos \alpha & -\sin \alpha \\ \sin \alpha & \cos \alpha \end{pmatrix}$$
$$= \begin{pmatrix} \cos \beta \cos \alpha - \sin \beta \sin \alpha & -(\cos \beta \sin \alpha + \sin \beta \cos \alpha) \\ \sin \beta \cos \alpha + \cos \beta \sin \alpha & -\sin \beta \sin \alpha + \cos \beta \cos \alpha \end{pmatrix}$$

hence

$$\cos(\alpha+\beta) = \cos \alpha \cos \beta - \sin \alpha \sin \beta \qquad (1)$$

and

$$\sin(\alpha+\beta) = \sin \alpha \cos \beta + \cos \alpha \sin \beta \qquad (2)$$

Note: This proof holds for α and β measured in radians or degrees.

AN ALTERNATIVE PROOF FOR
cos (α + β) = cos α cos β − sin α sin β

Consider a circle centre O and radius 1 unit.
  Let OA be the initial radius, hence A has coordinates (1, 0).
  Let A → B by a rotation of α radians about O, hence B has coordinates (cos α, sin α).
  Let B → C by a rotation of β radians about O, hence ∠AOC = α + β ∴ C has coordinates [cos(α + β), sin(α + β)].

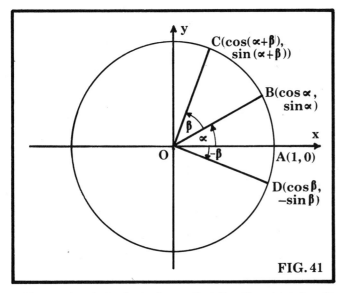

FIG. 41

Rotate the circle clockwise so that C → B and let A → D

$$\therefore \angle AOD = -\beta$$

and D has coordinates [cos(−β), sin(−β)], i.e. D has coordinates (cos β, −sin β).
  By this rotation CA → BD

$$\therefore CA^2 = BD^2$$

and this gives, by the distance formula,

$$[\cos(\alpha+\beta)-1]^2 + [\sin(\alpha+\beta)-0]^2$$
$$= (\cos\alpha - \cos\beta)^2 + (\sin\alpha + \sin\beta)^2$$

i.e.

$$\cos^2(\alpha+\beta) - 2\cos(\alpha+\beta) + 1 + \sin^2(\alpha+\beta)$$
$$= \cos^2\alpha - 2\cos\alpha\cos\beta + \cos^2\beta + \sin^2\alpha$$
$$+ 2\sin\alpha\sin\beta + \sin^2\beta$$

$$\therefore \cos^2(\alpha+\beta) + \sin^2(\alpha+\beta) + 1 - 2\cos(\alpha+\beta)$$
$$= (\cos^2\alpha + \sin^2\alpha) + (\cos^2\beta + \sin^2\beta)$$
$$- 2\cos\alpha\cos\beta + 2\sin\alpha\sin\beta$$

$$\therefore 2 - 2\cos(\alpha+\beta) = 2 - 2\cos\alpha\cos\beta + 2\sin\alpha\sin\beta$$
$$- 2\cos(\alpha+\beta) = -2\cos\alpha\cos\beta + 2\sin\alpha\sin\beta$$

$$\Rightarrow \quad \textbf{cos}\,(\boldsymbol{\alpha+\beta}) = \textbf{cos}\,\boldsymbol{\alpha}\,\textbf{cos}\,\boldsymbol{\beta} - \textbf{sin}\,\boldsymbol{\alpha}\,\textbf{sin}\,\boldsymbol{\beta} \quad (1)$$

and

$$\sin(\alpha+\beta) = \cos\left[\tfrac{1}{2}\pi - (\alpha+\beta)\right] = \cos\left[(\tfrac{1}{2}\pi - \alpha) - \beta\right]$$
$$= \cos\left[(\tfrac{1}{2}\pi - \alpha) + (-\beta)\right]$$
$$= \cos(\tfrac{1}{2}\pi - \alpha)\cos(-\beta) - \sin(\tfrac{1}{2}\pi - \alpha)\sin(-\beta)$$

$$\Rightarrow \quad \textbf{sin}\,(\boldsymbol{\alpha+\beta}) = \textbf{sin}\,\boldsymbol{\alpha}\,\textbf{cos}\,\boldsymbol{\beta} + \textbf{cos}\,\boldsymbol{\alpha}\,\textbf{sin}\,\boldsymbol{\beta} \quad (2)$$

Also

$$\cos(\alpha-\beta) = \cos[\alpha + (-\beta)]$$
$$= \cos\alpha\cos(-\beta) - \sin\alpha\sin(-\beta)$$

$$\Rightarrow \quad \textbf{cos}\,(\boldsymbol{\alpha-\beta}) = \textbf{cos}\,\boldsymbol{\alpha}\,\textbf{cos}\,\boldsymbol{\beta} + \textbf{sin}\,\boldsymbol{\alpha}\,\textbf{sin}\,\boldsymbol{\beta} \quad (3)$$

and

$$\sin(\alpha-\beta) = \sin[\alpha + (-\beta)]$$
$$= \sin\alpha\cos(-\beta) + \cos\alpha\sin(-\beta)$$

$$\Rightarrow \quad \textbf{sin}\,(\boldsymbol{\alpha-\beta}) = \textbf{sin}\,\boldsymbol{\alpha}\,\textbf{cos}\,\boldsymbol{\beta} - \textbf{cos}\,\boldsymbol{\alpha}\,\textbf{sin}\,\boldsymbol{\beta} \quad (4)$$

Example 1
Find the exact value of (i) sin 15° and (ii) cos 105°.

(i) $\sin 15° = \sin(45-30)°$
$= \sin 45° \cos 30° - \cos 45° \sin 30°$
$= \dfrac{1}{\sqrt{2}} \cdot \dfrac{\sqrt{3}}{2} - \dfrac{1}{\sqrt{2}} \cdot \dfrac{1}{2}$
$= \dfrac{\sqrt{3}-1}{2\sqrt{2}}$

(ii) $\cos 105° = \cos(60+45)°$
$= \cos 60° \cos 45° - \sin 60° \sin 45°$
$= \dfrac{1}{2} \cdot \dfrac{1}{\sqrt{2}} - \dfrac{\sqrt{3}}{2} \cdot \dfrac{1}{\sqrt{2}}$
$= \dfrac{1-\sqrt{3}}{2\sqrt{2}}$

Example 2
Without using tables evaluate $\cos 64° + \sin 64° \tan 32°$.

$\cos 64° + \sin 64° \tan 32° = \cos 64° + \sin 64° \dfrac{\sin 32°}{\cos 32°}$
$= \dfrac{\cos 64° \cos 32° + \sin 64° \sin 32°}{\cos 32°}$
$= \dfrac{\cos(64-32)°}{\cos 32°}$
$= \dfrac{\cos 32°}{\cos 32°} = 1$

Example 3
Find the value of $\cos(A-B)$, if $\tan A = \frac{3}{4}$ and $\tan B = \dfrac{1}{\sqrt{3}}$, and A and B are acute angles.

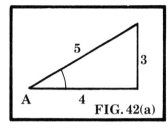
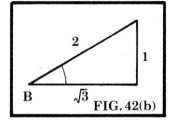

FIG. 42(a)   FIG. 42(b)

$\tan A = \dfrac{3}{4} \Rightarrow \sin A = \dfrac{3}{5}$ and $\cos A = \dfrac{4}{5}$

[figure 42(a)]

$\tan B = \dfrac{1}{\sqrt{3}} \Rightarrow \sin B = \dfrac{1}{2}$ and $\cos B = \dfrac{\sqrt{3}}{2}$

[figure 42(b)]

and

$\cos(A-B) = \cos A \cos B + \sin A \sin B$
$= \dfrac{4}{5} \cdot \dfrac{\sqrt{3}}{3} + \dfrac{3}{5} \cdot \dfrac{1}{2}$
$= \dfrac{4\sqrt{3}+3}{10}$, (the exact value)
$= \dfrac{4 \times 1 \cdot 73 + 3}{10} = \dfrac{6 \cdot 92 + 3}{10}$
$= 0 \cdot 992$ (an approximate value)

Example 4
Prove that $\cos\left(\dfrac{\pi}{2}+x\right) = -\sin x$.

$\cos\left(\dfrac{\pi}{2}+x\right) = \cos\dfrac{\pi}{2}\cos x - \sin\dfrac{\pi}{2}\sin x$
$= 0 - 1 \cdot \sin x$
$= -\sin x$

Example 5

Find the solution set of the equation $3 \sin \alpha° = 2 \sin(\alpha - 60)°$ for $0 \leq \alpha \leq 360$.

$$3 \sin \alpha° = 2 \sin(\alpha - 60)°$$
$$\Leftrightarrow 3 \sin \alpha° = 2(\sin \alpha° \cos 60° - \cos \alpha° \sin 60°)$$
$$\Leftrightarrow 3 \sin \alpha° = 2\left(\frac{1}{2} \sin \alpha° - \cos \alpha° \cdot \frac{\sqrt{3}}{2}\right)$$
$$\Leftrightarrow 3 \sin \alpha° = \sin \alpha° - \sqrt{3} \cos \alpha°$$
$$\Leftrightarrow 2 \sin \alpha° = -\sqrt{3} \cos \alpha°$$
$$\Rightarrow \tan \alpha° = -\frac{\sqrt{3}}{2} = -0·866$$
$$\Rightarrow \alpha = 180 - 40·9 \text{ or } 360 - 40·9$$

hence solution set is $\{139·1, 319·1\}$

ASSIGNMENT 2.3

1. Expand,
   (i) $\sin(x+y)$
   (ii) $\cos(a+b)°$
   (iii) $\cos(\theta - 2\alpha)$
   (iv) $\sin(2\theta - \beta)°$

2. (i) Verify the formula for $\cos(a-b)$ for $a = 60$ and $b = 30$.
   (ii) Verify the formula $\sin(a+b)°$ for $a = 60$ and $b = 30$.

3. Express as a single sine or cosine,
   (i) $\sin 40° \cos 15° + \cos 40° \sin 15°$
   (ii) $\cos 80° \cos 14° - \sin 80° \sin 14°$
   (iii) $\cos 2x \cos y + \sin 2x \sin y$
   (iv) $\sin 3\theta \cos \theta - \cos 3\theta \sin \theta$
   (v) $\sin 2\phi \cos(-\phi) - \cos 2\phi \sin(-\phi)$
   (vi) $\cos 5\beta° \cos \beta° + \sin 5\beta° \sin \beta°$.

4. Find the exact value of,
   (i) $\cos 15°$ (ii) $\cos 75°$.

In questions 5–9, prove that,

5. $\sin(90+A)° = \cos A°$.

6. $\cos(90+A)° = -\sin A°$.

7. $\sin 18° \cos 12° + \cos 18° \sin 12° = \frac{1}{2}$.

8. $\sin(A+30)° \cos(A+20)° - \cos(A+30)° \sin(A+20)°$ is independent of A.

9. $\sin(P-45)° = \frac{1}{\sqrt{2}}(\sin P° - \cos P°)$.

10. Find (without using tables) the value of,
    (i) $\cos 200° \cos 130° - \sin 200° \sin 130°$
    (ii) $\sin \frac{5\pi}{6} \cos \frac{\pi}{3} - \cos \frac{5\pi}{6} \sin \frac{\pi}{3}$.

11. Simplify,
    (i) $\cos(\frac{1}{3}\pi - \alpha) \cos \alpha - \sin(\frac{1}{3}\pi - \alpha) \sin \alpha$
    (ii) $\sin 220° \cos 140° + \cos 220° \sin 140°$ (without using tables)
    (iii) $\cos\left(\alpha + \frac{2\pi}{3}\right) + \cos \alpha + \cos\left(\alpha - \frac{2\pi}{3}\right)$.

12. Find (without using tables) the value of $\cos 42° + \sin 42° \tan 21°$.

13. If $\sin \alpha = \frac{1}{2}$ and $\cos \beta = \frac{1}{\sqrt{2}}$, and α and β are acute angles, find without using tables the value of,
    (i) $\sin(\alpha + \beta)$ (ii) $\cos(\alpha + \beta)$

14. If P and Q are acute angles such that $\tan P = \frac{4}{3}$ and $\tan Q = \frac{12}{5}$ find without using tables the value of,
    $\sin(P-Q)$ and $\cos(P-Q)$

15. Solve the following equations for the given intervals.
    (i) $\sin 60° \cos x° + \cos 60° \sin x° = 1, 0 \leq x \leq 180$
    (ii) $\cos 30° \cos x° - \sin 30° \sin x° = \frac{1}{2}, 0 \leq x \leq 180$
    (iii) $\cos\left(\frac{\pi}{6} + \alpha\right)\cos\left(\frac{\pi}{6} - \alpha\right) + \sin\left(\frac{\pi}{6} + \alpha\right)\sin\left(\frac{\pi}{6} - \alpha\right)$
    $= \frac{1}{2}, 0 \leq \alpha \leq \pi$
    (iv) $\sin(\frac{1}{3}\pi + \alpha) - \sin(\frac{1}{3}\pi - \alpha) = \frac{1}{2}, 0 < \alpha < 2\pi$.

16. Solve the equation, $3 \cos \alpha° = 2 \cos(\alpha - 60)°$, for $0 \leq \alpha \leq 360$.

17. Show that $\sin(45 + x)° + \sin(45 - x)° = \sqrt{2} \cos x°$.

18. ABC is a triangle with $AB = 5$, $BC = \sqrt{17}$ and $CA = 4$. Show, without using tables, that $\cos BAC = \frac{3}{5}$.
    If O is a point within the triangle ABC such that the angle $OAC = 45°$, show that,
    $$\cos OAB = \frac{7\sqrt{2}}{10}$$

19. In any triangle PQR, show that $\sin P = \sin(Q+R)$. If $\sin Q = \frac{1}{\sqrt{5}}$ and $\sin R = \frac{1}{\sqrt{10}}$, Q and R acute angles, find without using tables the value of $\sin P$.

20. If p and q are acute angles such that $\cos p = \frac{1}{\sqrt{5}}$ and $\sin q = \frac{4}{5}$ find without using tables the value of $\cos(p+q)$.

21. In figure 43, OP and OQ are perpendicular radii of a circle centre O. OTRS is a rectangle, and angle $POR = 2\theta$.
    Show that
    $$\text{angle ORQ} = \frac{\pi}{4} + \theta$$
    and
    $$\cos ORQ = \frac{\cos\theta - \sin\theta}{\sqrt{2}}$$

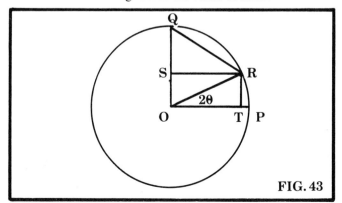

FIG. 43

FORMULAE FOR tan ($\alpha + \beta$) AND tan ($\alpha - \beta$) 2.5

$$\tan(\alpha+\beta) = \frac{\sin(\alpha+\beta)}{\cos(\alpha+\beta)} = \frac{\sin\alpha\cos\beta + \cos\alpha\sin\beta}{\cos\alpha\cos\beta - \sin\alpha\sin\beta}$$

$$= \frac{\dfrac{\sin\alpha\cos\beta}{\cos\alpha\cos\beta} + \dfrac{\cos\alpha\sin\beta}{\cos\alpha\cos\beta}}{\dfrac{\cos\alpha\cos\beta}{\cos\alpha\cos\beta} - \dfrac{\sin\alpha\sin\beta}{\cos\alpha\cos\beta}}$$

$$= \frac{\tan\alpha + \tan\beta}{1 - \tan\alpha\tan\beta} \qquad (5)$$

$$\tan(\alpha-\beta) = \frac{\sin(\alpha-\beta)}{\cos(\alpha-\beta)} = \frac{\sin\alpha\cos\beta - \cos\alpha\sin\beta}{\cos\alpha\cos\beta + \sin\alpha\sin\beta}$$

$$= \frac{\dfrac{\sin\alpha\cos\beta}{\cos\alpha\cos\beta} - \dfrac{\cos\alpha\sin\beta}{\cos\alpha\cos\beta}}{\dfrac{\cos\alpha\cos\beta}{\cos\alpha\cos\beta} + \dfrac{\sin\alpha\sin\beta}{\cos\alpha\cos\beta}}$$

$$= \frac{\tan\alpha - \tan\beta}{1 + \tan\alpha\tan\beta} \qquad (6)$$

Example 1
Find the exact value of tan 75°.

$$\tan 75° = \tan(45° + 30°) = \frac{\tan 45° + \tan 30°}{1 - \tan 45° \tan 30°}$$

$$= \frac{1 + 1/\sqrt{3}}{1 - 1 \cdot 1/\sqrt{3}}$$

$$= \frac{\sqrt{3} + 1}{\sqrt{3} - 1} \quad \text{and may be written}$$

$$= \frac{(\sqrt{3} + 1)(\sqrt{3} + 1)}{(\sqrt{3} - 1)(\sqrt{3} + 1)} = \frac{3 + 2\sqrt{3} + 1}{3 - 1}$$

$$= \frac{4 + 2\sqrt{3}}{2} = 2 + \sqrt{3}$$

Example 2
If $\tan \alpha = \frac{1}{3}$ and $\tan(\alpha + \beta) = 3$, find the value of $\tan \beta$.

$$\tan(\alpha + \beta) = \frac{\tan \alpha + \tan \beta}{1 - \tan \alpha \tan \beta}$$

$$\Rightarrow 3 = \frac{\frac{1}{3} + \tan \beta}{1 - \frac{1}{3}\tan \beta}$$

$$\Rightarrow 3 - \tan \beta = \frac{1}{3} + \tan \beta$$

$$\Rightarrow 2\frac{2}{3} = 2 \tan \beta$$

$$\Rightarrow \tan \beta = 1\frac{1}{3} = \frac{4}{3}$$

ASSIGNMENT 2.4

1. Expand,
   (i) $\tan(x + y)$ (ii) $\tan(2\alpha + \beta)$ (iii) $\tan(3\theta - \phi)$

2. Verify the formula for $\tan(\alpha + \beta)$ when $\alpha = \beta = \frac{\pi}{4}$.

3. Find without using tables the value of,
   (i) $\dfrac{\tan 40° - \tan 10°}{1 + \tan 40° \tan 10°}$ (ii) $\dfrac{\tan 26° + \tan 34°}{1 - \tan 26° \tan 34°}$

4. Find the exact value of tan 15° and tan 105°.

5. Prove that,
   (i) $\tan(a + 45)° = \dfrac{1 + \tan a°}{1 - \tan a°}$
   (ii) $\tan(a - 30)° = \dfrac{\sqrt{3} \tan a° - 1}{\sqrt{3} + \tan a°}$
   (iii) $\tan(\alpha + 60)° = \dfrac{\tan \alpha° + \sqrt{3}}{1 - \sqrt{3} \tan \alpha°}$

6. If $\sin p = \frac{4}{5}$ and $\cos q = \frac{15}{17}$, show without using tables that $\tan(p+q) = \frac{84}{13}$ and $\tan(p-q) = \frac{36}{77}$.

7. A window PR is 2 metres high and the base of the window is 5 metres from the ground which is horizontal. Q is a point on the ground $2\frac{1}{2}$ metres from the foot of the wall. Find the value of tan PQR.

8. If A and B are acute angles such that $\cos A = \frac{12}{13}$ and $A + B = 45°$ show without using tables that $\tan B = \frac{7}{17}$.

9. Given $\tan \alpha° = \dfrac{k}{k+1}$ and $\tan \beta° = \dfrac{1}{2k-1}$, $k \neq -1$ or $\frac{1}{2}$.
   Show that $\tan(\alpha + \beta)° = \dfrac{2k^2 + 1}{2k^2 - 1}$

10. Solve the equations for the interval $0 \leq x \leq 360$,
    (i) $\tan(45 - x)° = \tan x° - 1$
    (ii) $\tan(45 + x)° = \tan(45 - x)°$

11. Show that if angles P and Q are acute angles such that $\sin P = \frac{4}{5}$ and $\tan Q = \frac{1}{7}$, then $P - Q = \frac{\pi}{4}$.

12. In figure 44 PQ and RS are two vertical posts on a horizontal ground. PQ = 5 metres, RS = 8 metres and T is a point on the ground whose distance from the bases Q and S is $2\frac{1}{4}$ and $3\frac{1}{2}$ metres respectively. Find the value of tan PTR.

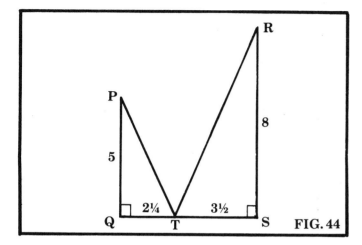

FIG. 44

ADDITION FORMULAE—PART (ii)
FORMULAE FOR THE DOUBLE ANGLES (i.e. 2α)

1. In $\sin(\alpha+\beta) = \sin\alpha\cos\beta + \cos\alpha\sin\beta$, replace β by α then $\sin(\alpha+\alpha) = \sin\alpha\cos\alpha + \cos\alpha\sin\alpha$

$$\therefore \sin 2\alpha = 2\sin\alpha\cos\alpha \qquad (1)$$

2. In $\cos(\alpha+\beta) = \cos\alpha\cos\beta - \sin\alpha\sin\beta$, replace β by α then $\cos(\alpha+\alpha) = \cos\alpha\cos\alpha - \sin\alpha\sin\alpha$

$$\therefore \cos 2\alpha = \cos^2\alpha - \sin^2\alpha \qquad (2a)$$
$$= \cos^2\alpha - (1-\cos^2\alpha)$$
$$= \cos^2\alpha - 1 + \cos^2\alpha$$
$$= 2\cos^2\alpha - 1 \qquad (2b)$$

also from
$$\cos 2\alpha = \cos^2\alpha - \sin^2\alpha$$
$$= 1 - \sin^2\alpha - \sin^2\alpha$$
$$= 1 - 2\sin^2\alpha \qquad (2c)$$

3. In
$$\tan(\alpha+\beta) = \frac{\tan\alpha + \tan\beta}{1 - \tan\alpha\tan\beta},$$

replace β by α then
$$\tan(\alpha+\alpha) = \frac{\tan\alpha + \tan\alpha}{1 - \tan\alpha\tan\alpha}$$

i.e. $$\tan 2\alpha = \frac{2\tan\alpha}{1 - \tan^2\alpha} \qquad (3)$$

Note: If we replace α by $\frac{1}{2}\theta$, it follows,
from formula (1) that $\sin\theta = 2\sin\frac{1}{2}\theta\cos\frac{1}{2}\theta$
from formula (2a) that $\cos\theta = \cos^2\frac{1}{2}\theta - \sin^2\frac{1}{2}\theta$
from formula (2b) that $\cos\theta = 2\cos^2\frac{1}{2}\theta - 1$
from formula (2c) that $\cos\theta = 1 - 2\sin^2\frac{1}{2}\theta$
from formula (3) that $\tan\theta = \dfrac{2\tan\frac{1}{2}\theta}{1 - \tan^2\frac{1}{2}\theta}.$

Also from formula (2b) $\cos^2\alpha = \frac{1}{2}(1+\cos 2\alpha)$
and from formula (2c) $\sin^2\alpha = \frac{1}{2}(1-\cos 2\alpha)$

SUMMARY OF ADDITION FORMULAE AND DOUBLE ANGLE FORMULAE

$$\sin(\alpha+\beta) = \sin\alpha\cos\beta + \cos\alpha\sin\beta$$
$$\sin(\alpha-\beta) = \sin\alpha\cos\beta - \cos\alpha\sin\beta$$
$$\cos(\alpha+\beta) = \cos\alpha\cos\beta - \sin\alpha\sin\beta$$
$$\cos(\alpha-\beta) = \cos\alpha\cos\beta + \sin\alpha\sin\beta$$
$$\tan(\alpha+\beta) = \frac{\tan\alpha + \tan\beta}{1 - \tan\alpha\tan\beta}$$

$$\tan(\alpha-\beta) = \frac{\tan\alpha - \tan\beta}{1+\tan\alpha\tan\beta}$$

$$\sin 2\alpha = 2\sin\alpha\cos\alpha$$

$$\cos 2\alpha = \cos^2\alpha - \sin^2\alpha$$
$$= 2\cos^2\alpha - 1$$
$$= 1 - 2\sin^2\alpha$$

$$\tan 2\alpha = \frac{2\tan\alpha}{1-\tan^2\alpha}$$

EQUATIONS

Example 1
Find the solution set of the equation
$$\cos 2\theta° + 3\sin\theta° - 2 = 0, \text{ for } 0 \leq \theta \leq 360.$$

$$\cos 2\theta° + 3\sin\theta° - 2 = 0$$
$$\Leftrightarrow 1 - 2\sin^2\theta° + 3\sin\theta° - 2 = 0$$
$$\Leftrightarrow -2\sin^2\theta° + 3\sin\theta° - 1 = 0$$
$$\Leftrightarrow 2\sin^2\theta° - 3\sin\theta° + 1 = 0$$
$$\Leftrightarrow (2\sin\theta° - 1)(\sin\theta° - 1) = 0$$
$$\Rightarrow \sin\theta° = \tfrac{1}{2} \text{ or } 1$$
$$\Rightarrow \theta = 30, 150, 90$$

Hence solution set is $\{30, 90, 150\}$.

Example 2
Solve the equation $3\sin 2x° = 2\sin x°$, where $0 \leq x < 360$.

$$3\sin 2x° = 2\sin x°$$
$$\Leftrightarrow 3.2\sin x°\cos x° = 2\sin x°$$
$$\Leftrightarrow 6\sin x°\cos x° - 2\sin x° = 0$$
$$\Leftrightarrow 2\sin x°(3\cos x° - 1) = 0$$
$$\Rightarrow \sin x° = 0 \text{ or } \cos x° = \tfrac{1}{3} = 0.333$$

$$\Rightarrow x = 0, 180 \text{ or } x = 70.55 \text{ or } 360 - 70.55$$

Hence solution set = $\{0, 70.55, 180, 289.45\}$.

Note: In solving equations involving double angles (i.e. 2α), we try (by using the formulae for the double angles) to reduce the given equation to,

(i) a quadratic in $\cos\alpha$ or $\sin\alpha$ or $\tan\alpha$ as in example 1.

or

(ii) an equation which has simple factors as in example 2.

Example 3
Find the maximum and minimum values of $\cos 2\theta + 2\cos\theta$.
Let
$$f(\theta) = \cos 2\theta + 2\cos\theta = 2\cos^2\theta - 1 + 2\cos\theta$$
$$= 2\cos^2\theta + 2\cos\theta - 1$$
$$= 2(\cos^2\theta + \cos\theta) - 1$$
$$= 2(\cos^2\theta + \cos\theta + \tfrac{1}{4}) - 1 - \tfrac{1}{2}$$
$$= 2(\cos\theta + \tfrac{1}{2})^2 - \tfrac{3}{2}$$

Now $(\cos\theta + \tfrac{1}{2})^2 \geq 0$ for all values of θ.

$\therefore$ the minimum value of $(\cos\theta + \tfrac{1}{2})^2 = 0$

Hence minimum value of $f(\theta) = -\tfrac{3}{2}$, occurring when $\cos\theta + \tfrac{1}{2} = 0$, i.e. when $\cos\theta = -\tfrac{1}{2}$, i.e. when $\theta = \dfrac{2\pi}{3}$ or $\dfrac{4\pi}{3}$ for $0 \leq \theta \leq 2\pi$.

Since the maximum value of $\cos\theta = 1$, then the maximum value of $f(\theta) = 2(1+\tfrac{1}{2})^2 - \tfrac{3}{2} = \tfrac{9}{2} - \tfrac{3}{2} = 3$ and occurs when $\cos\theta = 1$, i.e. $\theta = 0$ or 2π, for $0 \leq \theta \leq 2\pi$.

$\therefore$ We can write, for all values of θ,
$$-\tfrac{3}{2} \leq \cos 2\theta + 2\cos\theta \leq 3.$$

Example 4

Show that

$$\frac{\sin 2\theta}{1+\cos 2\theta} = \tan\theta$$

$$\text{L.H.S.} = \frac{\sin 2\theta}{1+\cos 2\theta} = \frac{2\sin\theta\cos\theta}{2\cos^2\theta}$$

(since $\cos 2\theta = 2\cos^2\theta - 1 \therefore 1+\cos 2\theta = 2\cos^2\theta$)

$$= \frac{\sin\theta}{\cos\theta}$$

$$= \tan\theta$$

ASSIGNMENT 2.5

1. Write down the formulae for $\sin 2\theta$, $\cos 2A$, $\tan 2B$.

2. Write down formulae for $\sin 4\theta$, $\cos 4x$, $\tan 4\phi$, in terms of 2θ, $2x$ and 2ϕ.

3. Write down 3 formulae for $\cos\theta$ in terms of $\frac{1}{2}\theta$.

4. If α is an acute angle find the value of $\sin 2\alpha$, $\cos 2\alpha$ and $\tan 2\alpha$ when,

    (i) $\sin\alpha = \frac{3}{5}$ (ii) $\cos\alpha = \frac{1}{2}$ (iii) $\tan\alpha = \frac{5}{12}$

5. Find the value of $\sin 2A$, $\cos 2A$ and $\tan 2A$, when,

    (i) $\tan A = \sqrt{3}$, and angle A acute
    (ii) $\tan A = \frac{8}{15}$, and angle A acute
    (iii) $\tan A = -1$ and A is an obtuse angle

6. Simplify,

    (i) $\cos^2 2\beta + \sin^2 2\beta$ (ii) $\frac{\sin 2\alpha}{\cos 2\alpha}$
    (iii) $2\sin 2\theta \cos 2\theta$ (iv) $2\cos^2 18° - 1$
    (v) $\cos^2 2x - \sin^2 2x$ (vi) $1+\cos 2A$
    (vii) $\sqrt{\left(\frac{1-\cos 2\alpha}{1+\cos 2\alpha}\right)}$

7. Find the value of,

    (i) $2\sin 15° \cos 15°$ (ii) $\cos^2 60° - \sin^2 60°$
    (iii) $\frac{2\tan 30°}{1-\tan^2 30°}$ (iv) $1 - 2\sin^2 15°$

8. Show that,

    (i) $(\cos\alpha - \sin\alpha)^2 = 1 - \sin 2\alpha$
    (ii) $(\cos\frac{1}{2}\beta + \sin\frac{1}{2}\beta)^2 = 1 + \sin\beta$

9. Find the exact value of $\dfrac{2\tan 22\cdot 5°}{1-\tan^2 22\cdot 5°}$ and hence show that the exact value of $\tan 22\cdot 5°$ can be written as $\sqrt{2}-1$.

10. Write down $\sin\theta\cos\theta$ in terms of 2θ. For what values of θ, $0 \leqq \theta < 2\pi$ will $\sin\theta\cos\theta$ have its maximum value? Write down this maximum value.

11. Write $\cos 4\theta$ in terms of $\cos 2\theta$ and $\cos 2\theta$ in terms of $\cos\theta$ and hence determine the constants p, q, r so that $\cos 4\theta = p\cos^4\theta + q\cos^2\theta + r$.

12. Write $\cos^2\theta$ in terms of $\cos 2\theta$ and hence express $\cos^4\theta$ in the form $a\cos 4\theta + b\cos 2\theta + c$ by determining the constants a, b, c.

13. By writing $\sin 3A$ as $\sin(A+2A)$ and using the formula for $\sin(\alpha+\beta)$ prove that $\sin 3A = 3\sin A - 4\sin^3 A$. Prove similarly that $\cos 3A = 4\cos^3 A - 3\cos A$.

14. Use the formulae obtained in question 13 to show that $\cos 6A = 32\cos^6 A - 48\cos^4 A + 18\cos^2 A - 1$.

15. Write $2\cos^2\alpha - \cos\alpha + 7$ in the form $a(\cos\alpha + b)^2 + c$ and show that $6\frac{7}{8} \leqq 2\cos^2\alpha - \cos\alpha + 7 \leqq 10$.

16. By writing $f(\theta) = \cos 2\theta° - \cos \theta°$ in the form $a(\cos \theta° + b)^2 + c$ show that $f(\theta)$ has a minimum value of $-1\frac{1}{8}$ occurring when $\theta = 75.5°$ and a maximum value of 2 occurring when $\theta = 180°$ $(0 \leq \theta \leq 180°)$.

17. By writing $\cos 2\theta - \sin \theta$ in the form $c - a(\sin \theta + b)^2$ show that $-2 \leq \cos 2\theta - \sin \theta \leq 1\frac{1}{8}$.

18. By using the methods of questions 16 and 17, find the maximum and minimum values of,
    (i) $3 \cos 2\theta - 4 \sin \theta$ (ii) $\cos \theta - \cos 2\theta$
    (iii) $3 \cos 2\theta - 4 \cos \theta$

In questions 19–24, prove that,

19. $(\sin \theta + \cos \theta)^2 - \sin 2\theta = 1$.

20. $1 - \left(\sin \dfrac{\theta}{2} - \cos \dfrac{\theta}{2}\right)^2 = \sin \theta$.

21. $\cos^4 \theta - \sin^4 \theta = \cos 2\theta$.

22. $\dfrac{\sin 4A}{1 + \cos 4A} = \tan 2A$

23. $\dfrac{2 \tan A}{1 + \tan^2 A} = \sin 2A$

24. $\dfrac{\sin 2A}{1 + \cos 2A} \cdot \dfrac{\cos A}{1 + \cos A} = \tan \dfrac{A}{2}$

25. Prove that,
$$\dfrac{1 - \tan^2 \theta}{1 + \tan^2 \theta} = \cos 2\theta$$

and show that
$$-1 \leq \dfrac{1 - \tan^2 \theta}{1 + \tan^2 \theta} \leq 1$$

for all values of θ.

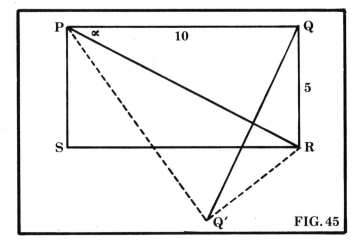

FIG. 45

26. In figure 45, PQRS is a rectangle with $PQ = 10$ units and $QR = 5$ units. Q' is the image of Q on reflection in PR.

    Using symmetry, state the lengths of RQ' and PQ'.
    If angle RPQ is denoted by α, express angle SPQ' in terms of α and show that $\cos SPQ' = \frac{4}{5}$.

Solve the following equations for the given intervals.

27. (i) $\sin 2\theta° = \frac{1}{2}$, $0 \leq \theta \leq 180$
    (ii) $\cos 2\theta° = -\dfrac{\sqrt{3}}{2}$, $0 \leq \theta \leq 180$
    (iii) $\tan 3\alpha = \sqrt{3}$, $0 \leq \alpha \leq \pi$.

28. (i) $\sin 2x° = \cos x°$, $0 \leq x \leq 360$
    (ii) $\sin 2x° + \sin x° = 0$, $0 \leq x < 360$
    (iii) $\tan 2x = 0$, $0 \leq x \leq 2\pi$.

29. Express $\cos 2x - \cos x = 0$ as a quadratic equation in $\cos x$ and hence solve the equation $\cos 2x - \cos x = 0$ for $0 \leq x \leq 2\pi$.

30. Solve the equation $1 + \sin x = \cos 2x$ for $0 \leq x \leq 2\pi$.

31. Solve the following equations,
    (i) $\tan 2\theta° + \tan \theta° = 0$, $0 \leq \theta \leq 360$
    (ii) $2\cos 2x° + \sin x° + 1 = 0$, $0 \leq x \leq 360$
    (iii) $2\cos^2 \theta = 1 + 2\sin\theta\cos\theta$, by writing it as an equation in $\tan 2\theta$, $0 \leq \theta \leq 2\pi$.

32. Find the solution set of the simultaneous equations,
    $$\sin\theta° + 2\cos\theta° = 1$$
    $$3\sin\theta° - \cos\theta° = 3, \quad 0 \leq \theta \leq 360$$

33. Solve for x the following equations,
    (i) $3\cos 4x° - 2 = \cos 2x°$, $0 \leq x \leq 180$
    (ii) $2\sin 4x° + \cos 2x° = 0$, $0 \leq x \leq 180$.

34. Find the solution set of the equation,
    $$\sin x°(2 - \sin^2 x°) = 2\sin 2x°, \quad 0 \leq x \leq 360$$

35. Solve the equation,
    $$3\cos 2x° + 2 + 13\sin x° = 0, \quad 0 \leq x \leq 360$$

ASSIGNMENT 2.6

Objective items testing Sections 2.3–2.6.
Instructions for answering these items are given on page 16.

1. If $\sin\theta = -\frac{1}{3}$ then the exact value of $\cos 2\theta$ is
    A. $\frac{1}{9}$
    B. $\frac{7}{9}$
    C. $\frac{1}{3}$
    D. $-\frac{7}{9}$
    E. $-\frac{1}{9}$

2. Triangle PQR is isosceles with $PR = PQ = 2$ units and angle PQR = angle $PRQ = \theta$. The length of the side p is
    A. $4\sin\theta$
    B. $4\cos\theta$
    C. $4\sin 2\theta$
    D. $4\cos 2\theta$
    E. It cannot be found from the information given.

3. If $\dfrac{\tan 2\theta}{\tan\theta} = p$ then an expression for $\tan^2\theta$ in terms of p would be
    A. $p(p+1)$
    B. $2 + \dfrac{1}{p}$
    C. $\dfrac{1}{2+p}$
    D. $1 - \dfrac{2}{p}$
    E. Some other expression involving p.

4. $\cos(\frac{2}{3}\pi - \theta)$ is equal to
    A. $-\frac{1}{2}(\cos\theta - \sqrt{3}\sin\theta)$
    B. $-\frac{1}{2}(\cos\theta + \sqrt{3}\sin\theta)$
    C. $\frac{1}{2}(\sqrt{3}\cos\theta - \sin\theta)$
    D. $\frac{1}{2}(\sqrt{3}\cos\theta + \sin\theta)$
    E. $-\frac{1}{2}(\sqrt{3}\cos\theta - \sin\theta)$

5. If $\cos \alpha \cos \beta = \dfrac{2-\sqrt{3}}{4}$ and $\sin \alpha \sin \beta = \dfrac{2+\sqrt{3}}{4}$ then for $0 \leq \alpha+\beta \leq \pi$, $\alpha+\beta$ equals

   A. $\dfrac{\pi}{6}$

   B. $\dfrac{\pi}{3}$

   C. π

   D. $\dfrac{2\pi}{3}$

   E. $\dfrac{5\pi}{6}$

6. $\dfrac{\cos \theta}{\sin \theta} - \dfrac{\cos 2\theta}{\sin 2\theta}$ when simplified is equal to

   A. $\dfrac{1}{\sin \theta}$

   B. $\dfrac{1}{\sin 2\theta}$

   C. $\tan \theta$

   D. $\dfrac{\sin 3\theta}{\sin 2\theta \sin \theta}$

   E. $\dfrac{\sin \theta}{\sin 3\theta}$

7. The least period of the function $f: x \to \cos 2x$, $x \in \mathbb{R}$, is

   A. $\dfrac{\pi}{4}$

   B. $\dfrac{\pi}{2}$

   C. π

   D. 2π

   E. 4π

8. If $q = \cos \theta$ and $p = \cos 2\theta$, then for all θ,

   A. $p = 1 - 2q^2$

   B. $p = 2q^2 - 1$

   C. $p^2 = 1 - q^2$

   D. $p^2 = 1 - 2q^2$

   E. $p^2 = 2q^2 - 1$

9. If $\tan(\alpha+\beta) = 2$ and $\tan \alpha = 3 \tan \beta$, then $\tan \beta$ has the value(s)

   (1) -1

   (2) $\tfrac{1}{3}$

   (3) 3

10. Which of the following angles belong(s) to the solution set of the equation

    $$\sin\left(\dfrac{\pi}{6}+\alpha\right)\cos\left(\dfrac{\pi}{6}-\alpha\right) - \cos\left(\dfrac{\pi}{6}+\alpha\right)\sin\left(\dfrac{\pi}{6}-\alpha\right) = \dfrac{1}{\sqrt{2}}$$

    for $0 \leq \alpha \leq 2\pi$?

    (1) $\dfrac{3\pi}{8}$

    (2) $\dfrac{9\pi}{8}$

    (3) $\dfrac{15\pi}{8}$

11. (1) $\cos 2\theta = -1$, $\theta \in \mathbb{R}$

    (2) $\theta = \dfrac{\pi}{2}$

12. (1) $\tan 2\theta = \dfrac{2h}{1-h^2}$

    (2) $h = \tan \theta$

UNIT 3: THE PRODUCTS AND SUMS OF SINES AND COSINES

SUMS AND PRODUCTS OF SINES AND COSINES 3.1

From $\sin(\alpha+\beta) = \sin\alpha\cos\beta + \cos\alpha\sin\beta$
and $\sin(\alpha-\beta) = \sin\alpha\cos\beta - \cos\alpha\sin\beta$

Add $\sin(\alpha+\beta) + \sin(\alpha-\beta) = 2\sin\alpha\cos\beta$
Subtract $\sin(\alpha+\beta) - \sin(\alpha-\beta) = 2\cos\alpha\sin\beta$

Let $\alpha + \beta = C$
and $\alpha - \beta = D$
then $2\alpha = C + D$ and $2\beta = C - D$

$$\therefore \alpha = \frac{C+D}{2} \quad \text{and} \quad \beta = \frac{C-D}{2}$$

Hence for $C > D$

$$\sin C + \sin D = 2 \sin \frac{C+D}{2} \cos \frac{C-D}{2}$$

and

$$\sin C - \sin D = 2 \cos \frac{C+D}{2} \sin \frac{C-D}{2}$$

Similarly $\cos(\alpha+\beta) = \cos\alpha\cos\beta - \sin\alpha\sin\beta$
$\cos(\alpha-\beta) = \cos\alpha\cos\beta + \sin\alpha\sin\beta$

Add $\cos(\alpha+\beta) + \cos(\alpha-\beta) = 2\cos\alpha\cos\beta$
Subtract $\cos(\alpha+\beta) - \cos(\alpha-\beta) = -2\sin\alpha\sin\beta$

Hence
$$\cos C + \cos D = 2\cos\frac{C+D}{2}\cos\frac{C-D}{2}$$

and
$$\cos C - \cos D = -2\sin\frac{C+D}{2}\sin\frac{C-D}{2}$$

Hence the four formula for the sums and products of sines and cosines are:

$$\sin C + \sin D = 2\sin\frac{C+D}{2}\cos\frac{C-D}{2}$$

$$\sin C - \sin D = 2\cos\frac{C+D}{2}\sin\frac{C-D}{2}$$

$$\cos C + \cos D = 2\cos\frac{C+D}{2}\cos\frac{C-D}{2}$$

$$\cos C - \cos D = -2\sin\frac{C+D}{2}\sin\frac{C-D}{2}$$

This last formula may be written,
$$\cos D - \cos C = 2\sin\frac{C+D}{2}\sin\frac{C-D}{2}, C > D$$

These formulae are used to express,
(i) the sum or difference of two sines or cosines in terms of products
(ii) the products of sines and cosines in terms of the sum or difference of two sines or cosines.

Example 1

Express (i) $\sin 36° + \sin 22°$ and (ii) $\cos 52° - \cos 36°$ as products of sines and cosines.

(i) $\sin C + \sin D = 2\sin\frac{C+D}{2}\cos\frac{C-D}{2}$

where $C = 36°, D = 22°$

$\therefore \sin 36° + \sin 22° = 2\sin 29°\cos 7°$

(ii) $\cos C - \cos D = -2\sin\frac{C+D}{2}\sin\frac{C-D}{2}$ with $C > D$,

where $C = 52°, D = 36°$

$\therefore \cos 52° - \cos 36° = -2\sin 44°\sin 8°$

Example 2

Express (i) $\sin 30°\cos 12°$ and (ii) $2\cos\frac{\pi}{3}\cos\frac{\pi}{6}$ as the sum or difference of two sines or cosines.

(i) From $\sin C + \sin D = 2\sin\frac{C+D}{2}\cos\frac{C-D}{2}$

Let

$\frac{C+D}{2} = 30°$ and $\frac{C-D}{2} = 12°$

then $C = 42°$ and $D = 18°$

$\therefore 2\sin 30°\cos 12° = \sin 42° + \sin 18°$
$\therefore \sin 30°\cos 12° = \frac{1}{2}(\sin 42° + \sin 18°)$

(ii) From $\cos C + \cos D = 2\cos\frac{C+D}{2}\cos\frac{C-D}{2}$

Let

$\frac{C+D}{2} = \frac{\pi}{3}$ and $\frac{C-D}{2} = \frac{\pi}{6}$

then $C = \dfrac{\pi}{2}$ and $D = \dfrac{\pi}{6}$

$$\therefore 2\cos\dfrac{\pi}{3}\cos\dfrac{\pi}{6} = \cos\dfrac{\pi}{2} + \cos\dfrac{\pi}{6}$$

Example 3

Express $\sin 2° \sin 47°$ as the sum or difference of two sines or cosines.

From

$$\cos C - \cos D = -2\sin\dfrac{C+D}{2}\sin\dfrac{C-D}{2}, \quad C > D$$

Let

$$\dfrac{C+D}{2} = 47° \quad \text{and} \quad \dfrac{C-D}{2} = 2°$$

then $C = 49°$ and $D = 45°$

$$\therefore \cos 49° - \cos 45° = -2\sin 47° \sin 2°$$
$$\therefore \sin 2° \sin 47° = -\tfrac{1}{2}(\cos 49° - \cos 45°)$$
$$= \tfrac{1}{2}(\cos 45° - \cos 49°)$$

Example 4

Find the maximum and minimum values of the expression $\sin(x+15)° \cos(x-35)°$ and the least corresponding value of $x (x > 0)$ for which they occur.

From

$$\sin C + \sin D = 2\sin\dfrac{C+D}{2}\sin\dfrac{C-D}{2}$$

Let

$$\dfrac{C+D}{2} = x+15 \quad \text{and} \quad \dfrac{C-D}{2} = x-35$$

then $C = 2x - 20$ and $D = 50$. Hence

$$\sin(x+15)° \cos(x-35)° = \tfrac{1}{2}[\sin(2x-20)° + \sin 50°]$$

Since $\sin 50° = 0.766$ is constant, the maximum or minimum value depends on $\sin(2x-20)°$ and $-1 \leq \sin(2x-20) \leq 1$.

Hence maximum value $= \tfrac{1}{2}(1 + 0.766) = 0.883$, and occurs when $\sin(2x-20)° = 1$.

But $\sin(2x-20)° = 1 \Rightarrow 2x - 20 = 90, 450, \ldots$
$$\Rightarrow 2x = 110, 470, \ldots$$
$$\Rightarrow x = 55, 235, \ldots$$

Hence for $x > 0$, the least value of x at which the maximum value occurs is 55.

The minimum value $= \tfrac{1}{2}(-1 + 0.766) = -0.117$, and occurs when $\sin(2x-20)° = -1$.

But $\sin(2x-20)° = -1 \Rightarrow 2x - 20 = 270, 630, \ldots$
$$\Rightarrow 2x = 290, 650, \ldots$$
$$\Rightarrow x = 145, 325, \ldots$$

Hence for $x > 0$, the least value of x at which the minimum value occurs is 145.

Thus the maximum value of the expression is 0.833 and occurs when $x = 55$.

The minimum value of the expression is -0.117 and occurs when $x = 145$.

ASSIGNMENT 3.1 (A)

Express the following as products of sines and cosines,

1. $\sin 28° + \sin 10°$
2. $\cos 13° + \cos 37°$
3. $\sin 29° - \sin 13°$
4. $\cos 18° - \cos 42°$
5. $\sin 105° - \sin 51°$
6. $\cos 56° - \cos 20°$
7. $\cos 29° + \cos 39°$
8. $\sin 410° - \sin 290°$
9. $\cos x° + \cos 3x°$
10. $\cos(\alpha - 2\beta) - \cos\alpha$
11. $\sin(x+y) - \sin x$
12. $\cos(x+h) - \cos x$
13. $\cos(x+\tfrac{1}{2}\pi) + \cos(x-\tfrac{1}{2}\pi)$
14. $\sin 181° + \sin 69°$

15. $\sin(x+h) - \sin x$
16. $\cos 3a° - \cos(-a°)$
17. $\sin \tfrac{2}{3}\phi - \sin \tfrac{1}{3}\phi$
18. $\sin(\alpha+\beta) + \sin(\alpha-\beta)$
19. $\cos(x-10)° - \cos(3x-12)°$
20. $\sin(\tfrac{1}{4}\pi - 2\theta) - \sin(\tfrac{1}{4}\pi - \theta)$
21. $\cos \dfrac{\pi}{8} + \cos \dfrac{3\pi}{8}$
22. $\cos 5\theta + \cos \theta$
23. $\sin 8A - \sin 2A$
24. $\cos 75° + \cos 45° + \cos 5°$

Find the exact value of the following,

25. $\sin 75° - \sin 15°$
26. $\cos 75° + \cos 45° - \cos 15°$
27. $\cos 105° + \cos 15°$
28. $\dfrac{\cos 24° - \cos 66°}{\sin 24° - \sin 66°}$
29. $\dfrac{\sin 81° - \sin 21°}{\cos 81° + \cos 21°}$

Prove that,

30. $\dfrac{\sin 4\theta - \sin 2\theta}{\cos 4\theta + \cos 2\theta} = \tan \theta$

31. $\dfrac{\cos 2\theta - \cos 5\theta}{\sin 2\theta + \sin 5\theta} = \tan \dfrac{3\theta}{2}$

32. If $p = \sin 3\alpha - \sin \alpha$ and $q = \cos 3\alpha - \cos \alpha$, express p and q as products of sines and cosines, and hence prove that,
   (i) $p^2 + q^2 = 2 - 2\cos 2\alpha$
   (ii) $\dfrac{p}{q} = -\dfrac{1}{\tan 2\alpha}$

33. If $u = \sin 3\theta + \sin \theta$ and $v = \cos 3\theta + \cos \theta$, express u and v as products of sines and cosines, and hence prove that,
   (i) $v^2 - u^2 = 4\cos 4\theta \cos^2 \theta$
   (ii) $vu = 2\sin 4\theta \cos^2 \theta$
   (iii) $\dfrac{v^2 - u^2}{vu} = \dfrac{1}{\tan 2\theta} - \tan 2\theta$

34. Prove that,
   (i) $\cos \alpha + \cos 3\alpha + \cos 5\alpha + \cos 7\alpha = 4\cos \alpha \cos 2\alpha \cos 4\alpha$
   (ii) $\sin \alpha + \sin 3\alpha + \sin 5\alpha + \sin 7\alpha = 4\cos \alpha \cos 2\alpha \sin 4\alpha$

35. Prove that,
   $(\sin 3\alpha - \sin \alpha)^2 (\sin 3\alpha + \sin \alpha) = \sin \alpha \sin^2 4\alpha$

ASSIGNMENT 3.1 (B)

Express the following as a sum or difference of sines or cosines.

1. $2 \sin 46° \cos 20°$
2. $2 \cos 32° \sin 12°$
3. $2 \cos 39° \cos 21°$
4. $2 \sin 29° \sin 36°$
5. $2 \sin \alpha° \cos \beta°, \alpha > \beta$
6. $2 \sin x \sin y$
7. $2 \cos A \cos B$
8. $2 \sin P \cos Q, P < Q$
9. $\sin 44° \cos 11°$
10. $\sin 11° \cos 44°$
11. $2 \cos(x+y) \cos(x-y)$
12. $2 \sin \dfrac{5\pi}{8} \cos \dfrac{2\pi}{8}$
13. $2 \cos \tfrac{1}{2}\theta \sin \tfrac{3}{2}\theta$
14. $2 \cos(\beta + \tfrac{1}{3}\pi) \cos(\beta - \tfrac{1}{3}\pi)$
15. $2 \sin(A+B) \cos(A-B)$
16. $2 \sin 7\theta \cos 3\theta$
17. $2 \sin(2\alpha+\beta) \sin(2\alpha-\beta)$
18. $2 \sin \dfrac{\alpha+\beta}{2} \cos \dfrac{\alpha-\beta}{2}$
19. $2 \sin \dfrac{\alpha+\beta}{2} \sin \dfrac{\alpha-\beta}{2}$

20. $2\cos(2P+Q)\cos(P-Q)$

21. $2\sin(24+\alpha)°\cos(24-\alpha)°$

22. $2\cos(x+y-z)\sin(x-y+z)$

Prove that,

23. $2\cos(\alpha+\tfrac{1}{2}\pi)\sin(\alpha-\tfrac{1}{2}\pi) = \sin 2\alpha$

24. $2\sin\left(\alpha+\dfrac{\pi}{4}\right)\sin\left(\alpha+\dfrac{3\pi}{4}\right) = \cos 2\alpha$

25. $2\sin(\alpha+2\pi)\cos(\alpha-2\pi) = \sin 2\alpha$

Express the following as the sum or difference of two sines or cosines,

26. $2\sin(P+3Q)\sin(2P+5Q)$

27. $2\cos(\alpha+2\beta)\sin(3\alpha+4\beta)$

28. $2\cos(5\alpha+3\beta)\sin(\alpha-\beta)$

29. Prove that,
$$4\sin x \sin\left(x+\dfrac{\pi}{3}\right)\sin\left(x+\dfrac{2\pi}{3}\right) = \sin 3x$$

30. By expressing each of the following products as a sum or difference of two sines or cosines, find the maximum and minimum value of each product.
  (i) $2\sin(x+45)°\cos(x-45)°$
  (ii) $2\cos(x+20)°\cos(x-30)°$
  (iii) $2\sin(x-15)°\cos(x+35)°$

31. Show that the expression $2\sin(x+12)°\cos(x-48)°$ has a maximum value of $1\cdot 866$ when $x = 63$.

32. Prove that for all values of θ,
$$-\tfrac{1}{2} \leq 2\cos(\theta+15)°\cos(\theta-45)° \leq 1\tfrac{1}{2}$$
Find also the values of θ in the range $0 \leq \theta \leq 360$ for which the maximum and minimum values occur.

33. Show without using tables that,
$$\sin 53°\cos 23° - \sin 25°\sin 11° + \cos 132°\cos 12° = 0$$

34. Express $\sin\alpha\cos 3\alpha$ as the difference of two sines and hence show that the sum of the first 6 terms of the series,
$$\sin\alpha\cos 3\alpha + \sin 2\alpha\cos 6\alpha + \sin 4\alpha\cos 12\alpha + \ldots,$$
is $\sin 63\alpha\cos 65\alpha$.

EQUATIONS 3.2

Since the function $f: x \to \sin x°$ is periodic with period of $360°$, it follows that,
$$\sin x° = \sin(x+360)° = \sin(x+2.360)°$$
$$= \sin(x+3.360)° = \ldots$$
and
$$\sin x° = \sin(x-360)° = \sin(x-2.360)°$$
$$= \sin(x-3.360)° = \ldots$$

This can be written
$$\sin x° = \sin(x+n.360)°, \quad n \in \mathbb{Z}$$

For the cosine function, $\cos x° = \cos(x+n.360)°$, $n \in \mathbb{Z}$ since the cosine function has a period of $360°$.

The tangent function has a period of $180°$, and similarly,
$$\tan x° = \tan(x+n.180)°, \quad n \in \mathbb{Z}$$

If the angle is measured in radians, say θ, these become
$$\sin\theta = \sin(\theta+2n\pi), \quad n \in \mathbb{Z}$$
$$\cos\theta = \cos(\theta+2n\pi), \quad n \in \mathbb{Z}$$
$$\tan\theta = \tan(\theta+n\pi), \quad n \in \mathbb{Z}$$

This is illustrated in examples 1, 2 and 3.

Example 1

Solve the equation $\sin x° = \frac{1}{2}$, in the cases,

(i) $0 \leq x \leq 360$ (ii) $x \in \mathbb{R}$

(i) $\sin x° = \frac{1}{2} \Rightarrow x = 30$ or 150
Thus for $0 \leq x \leq 360$ the solution set is $\{30, 150\}$.

(ii) $\sin x° = \frac{1}{2}$ then $x = 30$ or 150 for $0 \leq x \leq 360$, but if $x \in \mathbb{R}$ and $\sin x°$ has a period of $360°$, then
$$\sin(30 + n.360)° = \frac{1}{2}$$
and
$$\therefore x = 30 + n.360, \quad n \in \mathbb{Z}$$
Similarly
$$x = 150 + n.360, \quad n \in \mathbb{Z}$$

Hence for $x \in \mathbb{R}$ the solution set is
$$\{30 + n.360\} \cup \{150 + n.360\}, \quad n \in \mathbb{Z}$$

Example 2

Solve the equation $\cos \theta = -\frac{1}{2}$, for

(i) $0 \leq \theta \leq 2\pi$ (ii) $\theta \in \mathbb{R}$

(i) $\cos \theta = -\frac{1}{2} \Rightarrow \theta = \pi - \frac{\pi}{3}$ or $\pi + \frac{\pi}{3} = \frac{2\pi}{3}$ or $\frac{4\pi}{3}$

Thus for $0 \leq \theta \leq 2\pi$ the solution set is $\left\{\frac{2\pi}{3}, \frac{4\pi}{3}\right\}$.

(ii) Since the cosine function has a period of $360°$ or 2π, if the angle is measured in radians, then

$\cos \theta = -\frac{1}{2}, \quad \theta \in \mathbb{R} \Rightarrow \theta = \frac{2\pi}{3} + n.2\pi$ or $\frac{4\pi}{3} + n.2\pi$

$= \frac{2\pi}{3} + 2n\pi$ or $\frac{4\pi}{3} + 2n\pi$

Hence for $\theta \in \mathbb{R}$ the solution set is

$$\left\{\frac{2\pi}{3} + 2n\pi\right\} \cup \left\{\frac{4\pi}{3} + 2n\pi\right\}, \quad n \in \mathbb{Z}$$

Example 3

Solve $\tan x° = -1$, for

(i) $0 \leq x \leq 360$ (ii) $x \in \mathbb{R}$

(i) $\tan x° = -1 \Rightarrow x = 180 - 45$ or $x = 360 - 45$.
Thus for $0 \leq x \leq 360$ the solution set is $\{135, 315\}$.

(ii) $\tan x° = -1 \Rightarrow x = 135$, for $0 \leq x \leq 180$, but if $x \in \mathbb{R}$ and as $\tan x$ has a period of 180, then
$$\tan(135 + n.180)° = -1, \quad n \in \mathbb{Z}$$

Hence
$$x = 135 + n.180, \quad n \in \mathbb{Z}$$

and the solution set for $x \in \mathbb{R}$ may be written as
$$\{x : x = 135 + n.180, \quad n \in \mathbb{Z}\}$$

Note: If $\tan x = -1$ and x is measured in radians, then

(i) for $0 \leq x \leq 2\pi$,

$$x = \pi - \frac{\pi}{4} \text{ or } 2\pi - \frac{\pi}{4} = \frac{3\pi}{4} \text{ or } \frac{7\pi}{4}$$

and solution set is $\left\{\frac{3\pi}{4}, \frac{7\pi}{4}\right\}$.

(ii) for $x \in \mathbb{R}$, and since $\tan x$ has a period of π, then,

$$\tan\left(\frac{3\pi}{4} + n.\pi\right) = -1, \quad n \in \mathbb{Z}$$

Thus the solution set is $\left\{x : x = \frac{3\pi}{4} + n\pi, \quad n \in \mathbb{Z}\right\}$.

Example 4

Find the solution set of the equation
$$\sin 4x° - \sin 2x° = \cos 3x°, \quad \text{for } 0 \leq x \leq 360$$

Now
$$\sin 4x° - \sin 2x° = 2 \cos 3x° \sin x°$$

Hence
$$\sin 4x° - \sin 2x° - \cos 3x° = 0$$
$$\Leftrightarrow 2\cos 3x° \sin x° - \cos 3x° = 0$$
$$\Leftrightarrow \cos 3x°(2\sin x° - 1) = 0$$
$$\Rightarrow \cos 3x° = 0 \quad \text{or} \quad \sin x° = \tfrac{1}{2}$$
$$\cos 3x° = 0 \Rightarrow 3x = 90, 270, 450, 630, 810, 990$$
$$\Rightarrow x = 30, 90, 150, 210, 270, 330$$

and
$$\sin x° = \tfrac{1}{2} \Rightarrow x = 30 \text{ or } 150$$

hence the solution set is $\{30, 90, 150, 210, 270, 330\}$.

Example 5
Prove that,
$$\frac{\sin 5\theta + \sin 3\theta}{\cos 3\theta + \cos \theta} = 2\sin 2\theta$$

and hence show that the equation
$$\frac{\sin 5\theta + \sin 3\theta}{\cos 3\theta + \cos \theta} = 5\sin \theta$$

has no solution unless θ is a multiple of π.

$$\frac{\sin 5\theta + \sin 3\theta}{\cos 3\theta + \cos \theta} = \frac{2\sin 4\theta \cos \theta}{2\cos 2\theta \cos \theta} = \frac{\sin 4\theta}{\cos 2\theta}$$
$$= \frac{2\sin 2\theta \cos 2\theta}{\cos 2\theta}$$
$$= 2\sin 2\theta$$

hence the equation
$$\frac{\sin 5\theta + \sin 3\theta}{\cos 3\theta + \cos \theta} = 5\sin \theta$$

has a solution
$$2\sin 2\theta = 5\sin \theta$$
$$\Leftrightarrow 4\sin \theta \cos \theta - 5\sin \theta = 0$$
$$\Leftrightarrow \sin \theta(4\cos \theta - 5) = 0$$
$$\Leftrightarrow \sin \theta = 0 \quad \text{or} \quad \cos \theta = \tfrac{5}{4}$$

$\cos \theta = \tfrac{5}{4} > 0$ has no solution and $\sin \theta = 0 \Rightarrow \theta = 0, \pm\pi, \pm 2\pi, \pm 3\pi, \ldots = n\pi, n \in \mathbf{Z}$.

i.e. the equation has no solution unless θ is a multiple of π.

ASSIGNMENT 3.2

Solve the following equations for the intervals shown. (x in degrees, α in radians.)

1. $\sin x° = \dfrac{\sqrt{3}}{2}$, for (i) $0 \leqq x \leqq 360$ and (ii) $x \in \mathbf{R}$.

2. $\cos x° = \tfrac{1}{2}$, for $x \in \mathbf{R}$.

3. $\sin x° = -\tfrac{1}{2}$, for (i) $0 \leqq x \leqq 360$ and (ii) $x \in \mathbf{R}$.

4. $\tan x° = 1$, $x \in \mathbf{R}$.

5. $\sin \alpha = \dfrac{1}{\sqrt{2}}$, for (i) $0 \leqq \alpha \leqq 2\pi$ and (ii) $\alpha \in \mathbf{R}$.

6. $\tan \alpha = -\sqrt{3}$, $\alpha \in \mathbf{R}$.

7. $\tan x° = 2{\cdot}078$, $x \in \mathbf{R}$.

8. $\cos x° = -0{\cdot}804$, $x \in \mathbf{R}$.

9. $\sin \alpha = 1$, $\alpha \in \mathbf{R}$.

10. $\cos \alpha = -1$, $\alpha \in \mathbf{R}$.

11. $\tan \alpha = 0$, for (i) $0 \leqq \alpha < 2\pi$, and (ii) $\alpha \in \mathbf{R}$.

Solve the following equations for the intervals $0 \leqq x \leqq 360$ or $0 \leqq \theta \leqq 2\pi$.

12. $\cos 2x° = 0$
13. $\sin 3x° = \dfrac{\sqrt{3}}{2}$
14. $\tan 3x° = -1$
15. $\tan 2\theta = 0$
16. $\sin 2\theta = 1$
17. $\cos 3\theta = -\dfrac{1}{\sqrt{2}}$
18. $\sin \dfrac{\theta}{2} = \dfrac{1}{2}$
19. $\tan \dfrac{\theta}{2} = -\dfrac{1}{\sqrt{3}}$
20. $\sin 3x° - \sin x° = 0$
21. $\cos 3x° = \cos x°$
22. $\cos 3\theta + 2\cos \theta = 0$
23. $\sin 4x° - \sin 3x° + \sin x° = 0$
24. $\sin \theta = 2\sin \theta \sin 2\theta + \sin 3\theta$
25. $\cos 3\theta - \cos \theta - \sin 2\theta = 0$
26. $\sin 3x° + \sin x° = \cos x°$
27. $\cos 2x° = \cos x°$
28. $\cos x° + \cos 2x° + \cos 3x° + \cos 4x° = 0$
29. Prove that
$$\dfrac{\cos \theta - \cos 5\theta}{\sin 2\theta + \sin 4\theta} = 2\sin \theta$$
and hence show that if θ is not a multiple of π, then the equation
$$\dfrac{\cos \theta - \cos 5\theta}{\sin 2\theta + \sin 4\theta} = \sin 2\theta$$
has no solution.

30. If x is not a multiple of π, show that
$$\dfrac{\sin 4x}{\sin x - \sin 3x} = -2\cos x,$$
and hence find the solution set of the inequation
$$\dfrac{\sin 4x}{\sin x - \sin 3x} < -1,$$
given that $0 < x < \dfrac{\pi}{2}$.

ASSIGNMENT 3.3

Objective items testing Sections 3.1–3.2.
Instructions for answering these items are given on page 16.

1. $\cos 37° - \cos 15°$ is equal to
   A. $\cos 22°$
   B. $-\cos 22°$
   C. $2\sin 26° \sin 11°$
   D. $-2\sin 26° \sin 11°$
   E. $2\cos 26° \cos 11°$

2. $\sin \tfrac{1}{2}\theta - \sin \tfrac{3}{2}\theta$ is equal to
   A. $-\sin \theta$
   B. $-\cos \theta \sin \tfrac{1}{2}\theta$
   C. $-2\cos \tfrac{1}{2}\theta \sin \theta$
   D. $-2\cos \theta \sin \tfrac{1}{2}\theta$
   E. $2\cos \theta \sin \tfrac{1}{2}\theta$

3. $2\cos(x+y)\sin(x-y)$ expressed as the sum or difference of two sines or cosines is
   A. $\sin \dfrac{x+y}{2} - \sin \dfrac{x-y}{2}$
   B. $\cos x + \cos y$
   C. $\sin 2x - \sin 2y$
   D. $\cos 2x + \cos 2y$
   E. $\sin 2x + \sin 2y$

4. The minimum value of $2\cos(x+45)° \sin(x-45)°$, $x \in \mathbf{R}$ is
   A. -2
   B. -1
   C. 1
   D. 0
   E. 1

5. Which of the equivalences in the following chain of statements is false?
   $$\cos x° - \cos 3x° = \sin 4x°$$
   A. $\Leftrightarrow 2\sin 2x° \sin x° = 2\sin 2x° \cos 2x°$
   B. $\Leftrightarrow \sin x° = 1 - 2\sin^2 x$
   C. $\Leftrightarrow 2\sin^2 x° + \sin x° - 1 = 0$
   D. $\Leftrightarrow (2\sin x° - 1)(\sin x° + 1) = 0$
   E. $\Leftrightarrow \sin x° = \tfrac{1}{2}$ or $\sin x° = -1$

6. If $2\sin(x+15)° \cos(x-15)° = \tfrac{3}{2}$ and $0 \leqq x \leqq 180$, then x equals
   A. 0
   B. 45
   C. 90
   D. 180
   E. None of these.

7. If $m = \sin 4\alpha - \sin 2\alpha$ and $n = \cos 4\alpha + \cos 2\alpha$, then m/n is equal to
   A. $\tan \alpha$
   B. $\dfrac{1}{\tan \alpha}$
   C. $\tan 3\alpha$
   D. $-\tan \alpha$
   E. $-\dfrac{1}{\tan 3\alpha}$

8. If $0 \leqq x \leqq 360$, the number of distinct roots of the equation $2\sin^2 x - 3\sin x - 1 = 0$ is
   A. 0
   B. 1
   C. 2
   D. 3
   E. 4

9. The solution set of the equation
   $$2\cos(30+x)° \cos(30-x)° = \tfrac{1}{2}$$
   for $0 \leqq x \leqq 360$, contains which of the following values of x?
   (1) 90
   (2) 180
   (3) 450

10. Which of the following are members of the solution set of the equation $\tan x° = 1$, $x \in \mathbf{R}$?
    (1) 135°
    (2) 315°
    (3) 405°

11. (1) $\sin 2\alpha - \sin 2\beta = 0$
    (2) $\alpha + \beta = \dfrac{\pi}{2}$

12. (1) $\cos \alpha - \cos 3\alpha = \sin \alpha - \sin 3\alpha$
    (2) $\sin 2\alpha = -\cos 2\alpha$

UNIT 4: THE FUNCTION $a \cos x + b \sin x$

THE FUNCTIONS $f: x \to a \cos x° + b \sin x°$ with x IN DEGREES and $f: x \to a \cos x + b \sin x$ with x IN RADIANS

4.1

Functions of these forms occur frequently in applied mathematics, and it is very useful to be able to transfer the expressions $a \cos x° + b \sin x°$ and $a \cos x + b \sin x$ into some other containing only one trigonometrical ratio. This is illustrated in the following examples.

Example 1

Express $f(x) = 4 \cos x° - 3 \sin x°$ in the form $k \cos(x-\alpha)°$ by finding k and α with $k > 0$ and $0 \leq \alpha < 360$. Let

$$4 \cos x° - 3 \sin x° = k \cos(x-\alpha)°$$
$$= k \cos x° \cos \alpha° + k \sin x° \sin \alpha°$$

Equating coefficients of $\cos x°$ and $\sin x°$ we have

$$k \cos \alpha° = 4 \quad \text{and} \quad k \sin \alpha° = -3$$

and hence

$$k^2(\cos^2 \alpha° + \sin^2 \alpha°) = 16 + 9$$
$$k^2 = 25$$
$$k = 5, \text{ since } k > 0$$

Also $\dfrac{k \sin \alpha°}{k \cos \alpha°} = \dfrac{-3}{4}$, i.e. $\tan \alpha° = \dfrac{-3}{4}$, with $\alpha°$ in the 4th quadrant, since $k \sin \alpha° = -3$ (k is positive and $\therefore \sin \alpha°$ is negative), and $k \cos \alpha° = 4$ (k is positive and $\therefore \cos \alpha°$ is positive), and hence α must lie in the fourth quadrant

$$\therefore \tan\alpha° = -\tfrac{3}{4} \Rightarrow \alpha° = -36\cdot9° \text{ or } 323\cdot1°$$
$$\therefore f(x) = 5\cos(x+36\cdot9)° \text{ or } 5\cos(x-323\cdot1)°$$

Hence, in general,
$$a\cos x° + b\sin x° = k\cos(x-\alpha)°$$
where
$$k = \sqrt{(a^2+b^2)} \quad \text{and} \quad \tan\alpha° = \frac{b}{a}$$

Note: We can express $f(x)$ in the form $k\sin(x-\alpha)°$ by writing
$$4\cos x° - 3\sin x° = k\sin(x-\alpha)°$$
$$= k\sin x°\cos\alpha° - k\cos x°\sin\alpha°$$
$$\therefore -k\sin\alpha° = 4 \quad \text{and} \quad k\cos\alpha° = -3$$

and hence
$$k^2(\cos^2\alpha° + \sin^2\alpha°) = 9+16$$
$$k^2 = 25$$
$$k = 5 \quad (k>0)$$

and
$$\frac{k\sin\alpha°}{k\cos\alpha°} = \frac{-4}{-3} \Rightarrow \tan\alpha° = \frac{4}{3}$$

with α in the 3rd quadrant.
$$\Rightarrow \alpha° = 180° + 53\cdot1° = 233\cdot1°$$
$$\therefore f(x) = 5\sin(x-233\cdot1)°$$
or $\qquad = 5\sin(x+126\cdot9)°$
since $\qquad 233\cdot1° = -126\cdot9°$

Similarly we can write $f(x)$ in the form $k\cos(x+\alpha)°$ or $k\sin(x+\alpha)°$.

ASSIGNMENT 4.1

In this exercise $k > 0$, $R > 0$ and $0 \leq \alpha < 360$, unless otherwise stated; $\sin x° = \sin(x \text{ degrees})$, $\sin x = \sin(x \text{ radians})$.

1. Find k or R and α for each of the following pairs of equations,
   (i) $k\sin\alpha° = 3$, $k\cos\alpha° = 4$
   (ii) $k\cos\alpha° = 5$, $k\sin\alpha° = -12$
   (iii) $R\sin\alpha° = -8$, $R\cos\alpha° = 15$
   (iv) $R\sin\alpha° = 1$, $R\cos\alpha° = \sqrt{3}$
   (v) $k\cos\alpha° = -6$, $k\sin\alpha° = -8$
   (vi) $k\cos\alpha° = -2$, $k\sin\alpha° = 6$

2. Express each of the following in the form $k\cos(x-\alpha)°$.
   (i) $3\cos x° + 4\sin x°$
   (ii) $2\cos x° - \sin x°$
   (iii) $-4\cos x° + 2\sin x°$

3. Express each of the following in the form $R\sin(x-\alpha)°$.
   (i) $6\cos x° - 8\sin x°$
   (ii) $\cos x° + \sin x°$

4. If θ is in radian measure and $0 \leq \alpha < 2\pi$, express
   (i) $\sqrt{3}\cos\theta + \sin\theta$ in the form $k\cos(\theta-\alpha)$
   (ii) $\cos\theta - \sqrt{2}\sin\theta$ in the form $R\sin(\theta-\alpha)$
   (iii) $-\cos\theta - \sin\theta$ in the form $R\sin(\theta-\alpha)$

5. Express each of the following in terms of $k\cos(x-\alpha)°$, $0 \leq x < 360$, or $k\cos(x-\alpha)$ radians, $0 \leq \alpha < 2\pi$.
   (i) $\sqrt{6}\sin x° - \sqrt{2}\cos x°$
   (ii) $3\cos x° + 2\sin x°$
   (iii) $\cos x° - 2\cdot4\sin x°$

(iv) $\cos x - \sin x$

(v) $\sqrt{3} \sin x + \cos x$

MAXIMUM AND MINIMUM VALUES

Example 1

Find the maximum and minimum values of the function defined by $f(x) = 6 \cos x° + 8 \sin x°$. Find also the values of x in the domain $0 \leq x < 360$ for which the maximum and minimum values occur.

Write

$$f(x) = 6 \cos x° + 8 \sin x° = k \cos(x - \alpha)°$$
$$\Leftrightarrow 6 \cos x° + 8 \sin x° = k \cos x° \cos \alpha° + k \sin x° \sin \alpha°$$
$$\therefore k \cos \alpha° = 6 \quad \text{and} \quad k \sin \alpha° = 8$$

hence

$$k^2(\cos^2 \alpha° + \sin^2 \alpha°) = 36 + 64$$
$$k^2 = 100$$
$$k = 10, \quad \text{since } k > 0$$

and

$$\frac{k \sin \alpha°}{k \cos \alpha°} = \frac{8}{6},$$

i.e. $\tan \alpha° = 1.333$ and α is in the first quadrant.

$$\Rightarrow \alpha° = 53.1°$$

Hence

$$f(x) = 10 \cos(x - 53.1)°$$

But the maximum value of $\cos \theta = 1$ and the minimum value of $\cos \theta = -1$, i.e. for all values of θ, $-1 \leq \cos \theta \leq 1$.

Hence the **Maximum Value** of $f(x) = 10 \times 1 = 10$, and occurs when

$$\cos(x - 53.1)° = 1$$

i.e. when

$$(x - 53.1) = 0$$

i.e. when

$$x = 53.1, \quad \text{for } 0 \leq x < 360$$

The **Minimum Value** of $f(x) = 10 \times (-1) = -10$, and occurs when

$$\cos(x - 53.1)° = -1$$

i.e. when

$$(x - 53.1) = 180$$

i.e. when

$$x = 180 + 53.1$$
$$= 233.1, \quad \text{for } 0 \leq x < 360$$

ASSIGNMENT 4.2

1. Write down the maximum and minimum value of the functions defined by the given formulae, and the corresponding replacements for x in the domain $0 \leq x < 360$.

   (i) $f(x) = 3 \sin x°$
   (ii) $f(x) = \cos(x - 27)°$
   (iii) $f(x) = 5 \sin(x - 60)°$
   (iv) $f(x) = \sin 2x°$
   (v) $f(x) = 3 \cos(2x - 10)°$

2. Obtain the maximum and minimum values of the following functions defined by $y = f(x)$, and the corresponding replacements for x in the domain $0 \leq x < 360$.

   (i) $y = 6 \sin x° + 8 \cos x°$
   (ii) $y = \cos x° - \sin x°$
   (iii) $y = -8 \cos x° - 15 \sin x°$

3. Obtain the maximum and minimum values of the following functions defined by $y = f(\theta)$, and the corresponding replacements for θ in the domain $0 \leq \theta < 2\pi$, θ in radians.
   (i) $y = \sqrt{2} \sin \theta - \sqrt{2} \cos \theta$
   (ii) $y = \sqrt{3} \sin \theta + \cos \theta$

4. Show that for all real values of x, $-13 \leq 5 \sin x° + 12 \cos x° \leq 13$.

5. A point $P(x, y)$ moves such that $x = 3 - \sin \theta$, $y = 4 + \cos \theta$. Express OP^2 in terms of $\sin \theta$ and $\cos \theta$ and hence find the maximum and minimum values of the length OP.

6. The depth p cm of water at a certain fishing port t hours after midnight is given by the formula $p = 500(2 \cos 30t° + 5 \sin 30t°)$. By expressing p in the form $k \cos(x-\alpha)°$ find the times of high and low water in the twenty four hours from midnight.

7. In figure 46, ABCD is a rectangle inscribed in a circle centre O. Diameter $AC = d$ units and angle $AOB = 2\theta$ radians, where d is constant and θ varies.

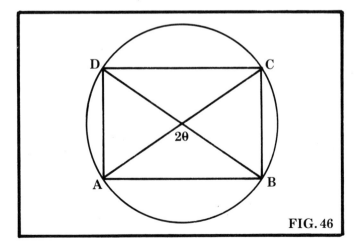

FIG. 46

Prove that the perimeter P units of the rectangle is given by $P = 2d(\cos \theta + \sin \theta)$.

By expressing P in the form $k \cos(\theta - \alpha)$, find in terms of d the maximum perimeter of the rectangle.

8. In figure 47, triangle PQR is right-angled at Q. QM is perpendicular to PR and angle $RQM = x$ radians. $PR = h$ units.

   Prove that the perimeter P of triangle PQR is given by $P = h + h(\cos x + \sin x)$.

   Prove that the maximum value of the perimeter is $h(1 + \sqrt{2})$ units.

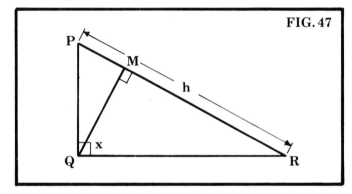

FIG. 47

SOLUTION OF EQUATIONS OF THE FORM
$a \cos x° + b \sin x° = c$

Example 1

Solve the equation $7 \cos x° + \sin x° = 5$, for $0 \leq x < 360$.

We first write the left hand side of the equation in the form $k \cos(x-\alpha)°$. Let

$$7 \cos x° + \sin x° = k \cos(x-\alpha)°$$
$$= k \cos x° \cos \alpha° + k \sin x° \sin \alpha°$$

then

$$k \cos \alpha° = 7 \quad \text{and} \quad k \sin \alpha° = 1$$

hence
$$k^2(\cos^2\alpha° + \sin^2\alpha°) = 49 + 1$$
$$k^2 = 50$$
$$k = \sqrt{50}, \quad k > 0$$

and
$$\tan\alpha° = \frac{k\sin\alpha°}{k\cos\alpha°} = \frac{1}{7} = 0\cdot143$$

with α in the first quadrant.
$$\Rightarrow \alpha° = 8\cdot1°$$

Hence
$$7\cos x° + \sin x° = \sqrt{50}\cos(x - 8\cdot1)°$$
$$\sqrt{50}\cos(x - 8\cdot1)° = 5$$
$$\Leftrightarrow \cos(x - 8\cdot1)° = \frac{5}{\sqrt{50}} = \frac{1}{\sqrt{2}}$$
$$\Rightarrow (x - 8\cdot1) = 45 \text{ or } 315$$
$$\Rightarrow x = 45 + 8\cdot1 \text{ or } 315 + 8\cdot1$$
$$= 53\cdot1 \text{ or } 323\cdot1$$

Example 2

Express $f(x) = 12\cos 2x° + 5\sin 2x°$ in the form $k\cos(2x - \alpha)°$ and hence solve the equation $12\cos 2x° + 5\sin 2x° = 2\cdot6$ for $0 \leq x < 360$.

Let
$$f(x) = 12\cos 2x° + 5\sin 2x°$$
$$= k\cos(2x - \alpha)°$$
$$= k\cos 2x° \cos\alpha° + k\sin 2x° \sin\alpha°$$

then
$$k\cos\alpha° = 12 \quad \text{and} \quad k\sin\alpha° = 5$$

hence
$$k^2(\cos^2\alpha° + \sin^2\alpha°) = 144 + 25$$
$$k^2 = 169$$
$$k = 13, \quad k > 0$$

and
$$\tan\alpha° = \frac{k\sin\alpha°}{k\cos\alpha°} = \frac{5}{12},$$

with α in the first quadrant.
$$\Rightarrow \alpha° = 22\cdot6°$$

Hence
$$f(x) = 12\cos 2x° + 5\sin 2x° = 13\cos(2x - 22\cdot6)°$$

and the equation is
$$13\cos(2x - 22\cdot6)° = 2\cdot6$$
$$\Rightarrow \cos(2x - 22\cdot6)° = \frac{2\cdot6}{13} = 0\cdot2$$
$$\Rightarrow (2x - 22\cdot6) = 78\cdot4 \text{ or } 281\cdot6 \text{ or } 438\cdot4 \text{ or } 641\cdot6$$
$$\Rightarrow 2x = 78\cdot4 + 22\cdot6 \text{ or } 281\cdot6 + 22\cdot6 \text{ or }$$
$$438\cdot4 + 22\cdot6 \text{ or } 641\cdot6 + 22\cdot6$$
$$\Rightarrow 2x = 101 \text{ or } 304\cdot2 \text{ or } 461 \text{ or } 664\cdot2$$
$$\Rightarrow x = 50\cdot5, 152\cdot1, 230\cdot5, 332\cdot1$$

Hence solution set is $\{50\cdot5, 152\cdot1, 230\cdot5, 332\cdot1\}$.

ASSIGNMENT 4.3

Solve the following equations for $0 \leq x < 360$.

1. $8\cos x° + 6\sin x° = 5$.

2. $5\sin x° - 12\cos x° = 2\cdot6$.

3. $\cos x° - 7\sin x° = 5$.

4. $4\cos x° + 3\sin x° = 2·5$.
5. $\sin x° + \cos x° + 1 = 0$.
6. $\sqrt{3}\cos x° + 2\sin x° = -0·7$.
7. $4\sin 2x° + 3\cos 2x° = 1·5$.
8. $1·2\cos 2x° - 0·9\sin 2x° = 0·5$.

Solve the following equations for $0 \leq \theta < 2\pi$, θ is in radian measure.

9. $\cos\theta + \sin\theta = -1$.
10. $\sqrt{3}\sin\theta - \cos\theta = 2$.
11. $\sqrt{3}\cos 2\theta - \sin 2\theta = \sqrt{3}$.
12. Express
$$f(x) = -7\cos x° + 7\sin x°$$
in the form $R\sin(x-\alpha)°$ and find the maximum and minimum values of the function $f(x)$. Solve the equation $f(x) = -\sqrt{2}$, $0 \leq x < 360$.

ASSIGNMENT 4.4

Objective items testing Sections 4.1–4.3.
Instructions for answering these items are given on page 16.

1. When $\cos x° - \sqrt{3}\sin x°$ is expressed in the form $R\cos(x-\alpha)°$ where $R > 0$ and $0 \leq \alpha < 360$ then
   A. $R = 2$ and $\alpha = 30$
   B. $R = 2$ and $\alpha = 60$
   C. $R = \sqrt{2}$ and $\alpha = 120$
   D. $R = \sqrt{2}$ and $\alpha = 240$
   E. $R = 2$ and $\alpha = 300$

2. When $\cos x + \sin x$ is expressed in the form $R\sin(x-\alpha)$ where $R > 0$ and $0 \leq \alpha < 2\pi$ then
   A. $R = \sqrt{2}$ and $\alpha = \dfrac{\pi}{4}$
   B. $R = \sqrt{2}$ and $\alpha = \dfrac{3\pi}{4}$
   C. $R = 1$ and $\alpha = \dfrac{5\pi}{4}$
   D. $R = \sqrt{2}$ and $\alpha = \dfrac{7\pi}{4}$
   E. $R = \sqrt{2}$ and $\alpha = \dfrac{5\pi}{4}$

3. When $\sqrt{3}\cos 2x - 3\sin 2x$ is expressed in the form $R\cos(2x-\alpha)°$ where $R > 0$ and $0 \leq \alpha < 360$, then α equals
   A. 60 only
   B. 120 only
   C. 300 only
   D. 330 only
   E. α has more than one value.

4. For $x \in R$ the maximum value of $5\sin x - 12\cos x$ is
   A. 3
   B. 5
   C. 12
   D. 13
   E. 15

5. If $\sqrt{3}\cos\theta + \sin\theta = 2\cos\left(\theta - \dfrac{\pi}{6}\right)$, the solution set of the equation $\sqrt{3}\cos\theta + \sin\theta = \sqrt{3}$, $0 \leq \theta < 2\pi$ is

   A. $\{0\}$
   B. $\left\{\dfrac{\pi}{3}\right\}$
   C. $\left\{0, \dfrac{\pi}{3}\right\}$
   D. $\left\{\dfrac{\pi}{3}, 2\pi\right\}$
   E. $\left\{0, \dfrac{\pi}{3}, 2\pi\right\}$

6. The minimum value of $\tfrac{1}{2}\sin x° - \tfrac{1}{2}\cos x°$ for $0 \leq x < 360$ occurs when

   A. $x = 45$ only
   B. $x = 45$ or 225
   C. $x = 315$ only
   D. $x = 135$ or 315
   E. $x = 225$ only

7. $f: x \to \sin 2x$. Which one of the following statements is false?

   A. f is a periodic function.
   B. The maximum value of $f(x)$ is 2.
   C. $f(x) = 0$ when $x = \dfrac{\pi}{2}$
   D. $f(x)$ can have the value -1.
   E. $\sin 2x = \cos 2x$ for $x = \dfrac{5\pi}{8}$

8. A function f is defined by $f: x \to \sqrt{3}\cos x + \sin x$, $x \in R$. When it is written in the form $k\cos(x - \alpha)$, which one of the following statements is false?

   A. $\alpha = \dfrac{\pi}{6}$ for $0 \leq \alpha < 2\pi$
   B. The maximum value of $f(x)$ is 2.
   C. k has the value 2.
   D. The equation $f(x) = 0$ gives a solution $x = \dfrac{\pi}{4}$.
   E. The minimum value of $f(x)$ is -2.

9. Which of the following are possible values of α when $\sin x° - \cos x°$ is written in the form $k\cos(x - \alpha)°$, $-360 \leq \alpha < 360$?

   (1) 135
   (2) -225
   (3) -135

10. If $f(\theta) = 4\cos\theta - 3\sin\theta$ is written in the form $R\sin(\theta - \alpha)$, $0 \leq \alpha < 2\pi$, the solution set of the equation $f(\theta) = 0$ will contain

    (1) $2\pi - \alpha$
    (2) $\pi + \alpha$
    (3) α

11. (1) $f(\theta) = \cos\theta° - \sin\theta°$
    (2) $f(\theta) = \sqrt{2}\cos(\theta - 45)°$

12. (1) $f(\theta) = \sqrt{3}\sin(\theta - 60)°$
    (2) $f(\theta) = \dfrac{\sqrt{3}}{2}\sin\theta° + \dfrac{\sqrt{3}}{2}\cos\theta°$

ASSIGNMENT 4.5

SUPPLEMENTARY EXAMPLES

1. Write in terms of an acute angle with the appropriate sign,
   - (i) $\sin 210°$
   - (ii) $\cos 136°$
   - (iii) $\tan 310°$
   - (iv) $\tan 120°$
   - (v) $\cos 316°$
   - (vi) $\sin 240°$
   - (vii) $\sin 139°$.

2. Write in terms of a positive acute angle,
   - (i) $\sin(-30°)$
   - (ii) $\tan(-120°)$
   - (iii) $\cos(-39°)$
   - (iv) $\cos(-129°)$
   - (v) $\tan(-300°)$
   - (vi) $\cos(-140°)$
   - (vii) $\sin(-210°)$.

3. If $\sin A = \frac{9}{41}$, calculate $\cos A$ and $\tan A$ when,
   - (i) angle A is acute
   - (ii) angle A is obtuse.

4. If $\tan A = \dfrac{p^2 - q^2}{2pq}$, find $\sin A$ and $\cos A$ in terms of p and q.

5. If $\tan P = \sqrt{3}$ and $\tan Q = \dfrac{1}{\sqrt{3}}$, express $\dfrac{\tan P - \tan Q}{1 + \tan P \tan Q}$ in its simplest form.

6. If $\cos x° = -\frac{3}{5}$ and $90 < x < 180$, find the value of $\sin x°$ and $\tan x°$.

7. Prove that,
   - (i) $\cos(90-A)° \sin(180-A)° = \sin^2 A°$
   - (ii) $\sin B \tan B = \dfrac{1-\cos^2 B}{\cos B}$
   - (iii) $\dfrac{1}{\tan^2 A} + 1 = \dfrac{1}{\sin^2 A}$
   - (iv) $\sin^4 A - \cos^4 A = \sin^2 A - \cos^2 A$
   - (v) $\cos(180-A)° \sin(90-A)° = -\cos^2 A°$
   - (vi) $\dfrac{\sin A - 2\sin^3 A}{2\cos^3 A - \cos A} = \tan A$
   - (vii) $\dfrac{1 - \tan^2 \theta}{1 + \tan^2 \theta} = 2\cos^2 \theta - 1$.

8. By using the exact values of sin, cos and tan, prove that
   - (i) $\sin 60° \cos 30° - \cos 60° \sin 30° = 0$
   - (ii) $\dfrac{\tan 60° - \tan 30°}{1 + \tan 60° \tan 30°} = \dfrac{1}{\sqrt{3}}$
   - (iii) $3\cos 30° - 4\cos^3 30° = 0$.

9. Find in simplest form the exact value of,
   - (i) $\sin^2 30° + \cos^2 30° + \tan^2 45°$
   - (ii) $\sin 60° \cdot \cos 60° \cdot \tan 60°$
   - (iii) $\sin 45° + \cos 45° + \tan 45°$.

10. Solve the following equations for $0 \leq \theta \leq 180$,
    - (i) $4\sin^2 \theta = 3$
    - (ii) $4\cos^2 \theta = 1$
    - (iii) $\cos^2 \theta - 3\cos \theta - 4 = 0$
    - (iv) $2\sin \theta - \cos^2 \theta = \frac{1}{4}$
    - (v) $1 + \tan^2 \theta = 3\tan \theta - 1$
    - (vi) $2\cos^2 \theta + 3\cos \theta - 5 = 0$
    - (vii) $3\cos^2 \theta - 4\cos \theta + 1 = 0$
    - (viii) $3\tan \theta = \tan^3 \theta$.

11. If $\tan A = -\frac{5}{12}$ and $90 < A < 180$, find the value of $\sin A$ and $\cos A$.

12. Write down in terms of A,
    - (i) $\cos(360-A)°$
    - (ii) $\tan(180+A)°$
    - (iii) $\sin(180-A)°$
    - (iv) $\sin(360-A)°$
    - (v) $\tan(180-A)°$
    - (vi) $\cos(180+A)°$.

13. Find the exact value of
    (i) tan 330° (ii) cos 210° (iii) cos 315° (iv) sin 120°
    (v) tan 225° (vi) tan 150° (vii) sin 300° (viii) cos 180°
    (ix) sin 240° (x) cos 135°.

14. Find the solution sets of the equations, for $0 \leq x < 360$,
    (i) $\sin x° = \dfrac{1}{\sqrt{2}}$
    (ii) $\cos x° = -\frac{1}{2}$
    (iii) $\tan x° = -\sqrt{3}$
    (iv) $\sin x° = -1$
    (v) $\cos x° = \dfrac{\sqrt{3}}{2}$
    (vi) $\tan x° = 1$
    (vii) $\tan x° = 1 \cdot 937$
    (viii) $\sin x° = -0 \cdot 819$
    (ix) $\cos x° = \sin x°$.

15. Find the solution sets of the equations for the interval $0 \leq \theta < 360$,
    (i) $3 \sin \theta° (\sin \theta° + 1) = 2 \cos^2 \theta°$
    (ii) $3 \tan^2 \theta° = 1 + 2 \tan \theta°$.

16. Figure 48 shows a tower CD, d metres high. A and B are two points from C the foot of the tower such that AB = BC. The angles of elevation of the top of the tower from A and B are $a°$ and $b°$ respectively.
    (i) Prove that $\tan b° = 2 \tan a°$.

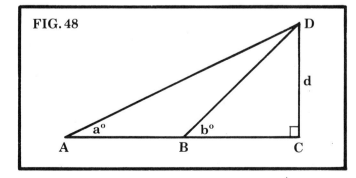

FIG. 48

(ii) If $b = 23$ and the tower is 70 metres high, calculate the value of a and the length of AB.

17. A and B are two stations 1000 metres apart on a straight stretch of sea-shore and B is due east of A. At A a lighthouse bears 035° and at B the lighthouse bears 318°. Show that the distance d metres of the lighthouse from the shore is given by the equation,

$$1000 = \dfrac{d}{\tan 55°} + \dfrac{d}{\tan 48°}$$

and hence find the distance of the lighthouse from the shore.

18. In figure 49, AC is a diameter of a circle of radius r. B is a point on the circumference of the circle such that BA = BC and X is the point on AC produced such that angle CBX = $\theta°$. Prove that,
    (i) angle BXC = $(45 - \theta)°$
    (ii) by applying the Sine Rule to triangle CBX,

    $$BX = \dfrac{r\sqrt{2}}{\cos \theta° - \sin \theta°}.$$

    If Y is on AC such that angle YBC = θ, prove that,
    (iii) $BY = \dfrac{r\sqrt{2}}{\cos \theta° + \sin \theta°}$
    (iv) area of triangle XBY = $r^2 \tan 2\theta°$.

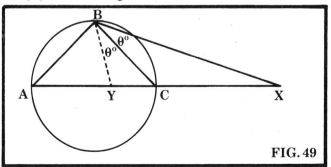

FIG. 49

19. In figure 50, ABC is an isosceles triangle with CA = CB = x units, inscribed in a circle centre O and radius r. The tangents at A and B meet at the point T.

If angle AOB = 2θ, prove that by applying the Cosine Rule to triangles AOB and ACB,

$$\frac{r^2}{x^2} = \frac{1-\cos\theta}{1-\cos 2\theta} = \frac{\sin^2 \tfrac{1}{2}\theta}{\sin^2 \theta}$$

and deduce that $x = 2r\cos\dfrac{\theta}{2}$. Similarly by applying the Cosine Rule to triangles AOB and ATB, or otherwise show that the tangents AT and BT have length $r\tan\theta$.

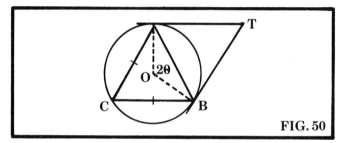

FIG. 50

20. A triangle ABC has AB = 4·9 cm, CB = 9·2 cm and has area 12·6 cm².
    (i) Calculate the two possible values of the angle B.
    (ii) Calculate the greater of the two possible lengths of AC.

21. From a station P a road runs south east and a man observes a tower S 62° E of him. After walking 1000 metres along the road he observes the tower again which is now S 85° E of him. Find the distance of the tower from the station and its shortest distance from the road.

22. (i) Express in terms of π the radian measure of, 15°, 40°, 60°, 210°, 270°, 300°, 360°.
    (ii) Express in degree measure, $\tfrac{1}{8}\pi$, $\tfrac{1}{6}\pi$, $\tfrac{2}{5}\pi$, $\tfrac{3}{4}\pi$, π, $\tfrac{2}{3}\pi$, $\tfrac{5}{6}\pi$, 2π radians.

23. The arc AB of a circle of radius 10 cm subtends an angle of 42° at O the centre of the circle. Calculate the length of the arc AB and the area of the sector AOB.

24. Assuming that the radius of the earth is 6600 kilometres, find the distance on the surface between two places on the same meridian, the difference of whose latitudes is 21°.

25. A circular wire of radius 15 cm is cut and bent to lie along the circumference of a wheel whose radius is 24 cm. Calculate the angle the wire subtends at the centre of the wheel and the area of the smaller sector enclosed by the wire and the radii.

26. Expand,
    (i) $\sin(2a+b)°$ (ii) $\cos(2a-b)°$
    (iii) $\sin(a-b)°$ (iv) $\cos(a+b)°$
    and verify each formula when $a = 30$ and $b = 60$.

27. Expand,
    (i) $\cos(\alpha-\beta)$ (ii) $\sin(2\alpha-\beta)$
    (iii) $\cos(\alpha+\beta)$ (iv) $\cos(2\alpha+\beta)$
    and verify each formula when $\alpha = \tfrac{1}{4}\pi$ and $\beta = \tfrac{1}{4}\pi$.

28. Expand,
    (i) $\tan(a-b)°$ (ii) $\tan(2\theta+\alpha)$
    and verify the formulae when $a = 60$ and $b = 30$ and when $\theta = \tfrac{1}{2}\pi$ and $\alpha = \tfrac{1}{4}\pi$.

29. Prove that,
    (i) $\sin 7° \cos 23° + \cos 7° \sin 23° = \tfrac{1}{2}$
    (ii) $\cos(180-a)° = -\cos a°$
    (iii) $\sin(90+a)° = \cos a°$.

30. Find the value of $\sin 2A$, $\cos 2A$ and $\tan 2A$ when angle A is acute and
    (i) $\sin A = \frac{1}{2}$ (ii) $\cos A \frac{3}{5}$ (iii) $\tan A = \frac{1}{3}$

31. P is the point with coordinates $(\cos \alpha, \sin \alpha)$ and Q the point with coordinates $(\cos \beta, \sin \beta)$. Show that the length of PQ is given by $2 \sin \frac{1}{2}(\alpha - \beta)$.

32. Solve the equations for $0 \leq x \leq 360$.
    (i) $3 \cos 2x° + 2 = \cos x°$
    (ii) $\sin 2x° + \cos x° = 0$
    (iii) $2 \cos 2x° + 3 \cos x° + 1 = 0$
    (iv) $3 \sin x° = 2 \sin (x-60)°$.

33. Solve the equations for $0 \leq \theta < 2\pi$.
    (i) $\sin 2\theta + \sin \theta = 0$ (ii) $\cos 2\theta + \sin \theta - 1 = 0$
    (iii) $\tan 2\theta + \tan \theta = 0$.

34. Express as the product of two sines or cosines,
    (i) $\sin 56° + \sin 32°$ (ii) $\cos 32° - \cos 56°$
    (iii) $\sin \frac{2}{3}\pi - \sin \frac{1}{3}\pi$ (iv) $\cos \frac{5}{2}\pi + \cos \frac{3}{2}\pi$

35. Express each of the following as the sum or difference of sines or cosines,
    (i) $2 \sin 60° \cos 30°$ (ii) $2 \cos 43° \cos 53°$
    (iii) $\sin 25° \cos 55°$ (iv) $\sin 50° \sin 70°$

36. Solve the following equations for the interval $0 \leq x < 360$.
    (i) $2 \cos^2 x° - \cos 3x° = \cos x°$
    (ii) $\sin x° - \cos 2x° - \sin 3x° = 0$
    (iii) $\cos 3x° + \cos x° = \cos 2x°$.

37. By writing $a \cos x + b \sin x$ in the form $R \cos (x - \alpha)$ solve the equations,
    (i) $7 \cos x° + \sin x° = 5, 0 \leq x < 360$
    (ii) $3 \sin x° + 4 \cos x° + 1 = 0, 0 \leq x < 360$
    (iii) $24 \sin x° - 7 \cos x° = 25, 0 \leq x < 360$.

38. Find the sum to infinity of the following series and state the range of values of θ in the interval $0 \leq \theta < \frac{1}{2}\pi$ for which the result is valid.
    (i) $\sin 2\theta + 2 \sin 2\theta \sin^2 \theta + 4 \sin 2\theta \sin^4 \theta + \cdots$
    (ii) $\cos 2\theta + 2 \cos 2\theta \cos^2 \theta + 4 \cos 2\theta \cos^4 \theta + \cdots$.

CALCULUS

UNIT 1: THE DIFFERENTIAL CALCULUS

NOTATION FOR FUNCTIONS 1.1

Let S and T be two non-empty sets. A relation f from S to T is called a **mapping** or **function** from S into T if

(i) the domain of f is the entire set S; and
(ii) each element of S is related by f to exactly one element of T.

If $a \in S$ and b is the single element of T to which a is related by f, then b is called the **image of a under the function f**, and it is customary to write $f(a) = b$.

A method of specifying a function f which is often used is to give the domain of f and a formula or rule for finding the image of each element of the domain.

Example 1

The function $f: x \rightarrow x^2 + 2$ maps the set R into the set R. This can also be written as "The function $f: R \rightarrow R$ defined by $f(x) = x^2 + 2$." Another notation is to write $y = x^2 + 2$ where y represents $f(x)$. i.e. $y = f(x) = x^2 + 2$.

With this notation x is called the **independent variable** and y the **dependent variable**.

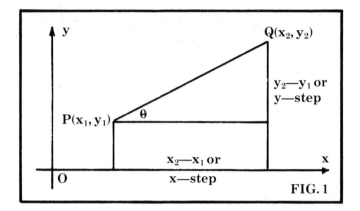

FIG. 1

1.2 GRADIENTS

The **gradient** of a straight line is the tangent of the angle the line makes with the positive direction of the x-axis.

In figure 1, P is the point (x_1, y_1) and Q the point (x_2, y_2) and the gradient of PQ, written

$$m_{PQ} = \frac{y_2 - y_1}{x_2 - x_1}, \quad x_2 \neq x_1.$$

The gradient of a straight line tells us the rate at which y increases (or decreases) for unit increase (or decrease) in x.

$$\text{Gradient of PQ} = \frac{y_2 - y_1}{x_2 - x_1}$$
$$= \frac{\text{increment in } y}{\text{increment in } x}$$
$$= \frac{y\text{-step}}{x\text{-step}}$$

Example 1

Find the gradient of the line joining the points $P(3, -2)$ and $Q(5, -8)$.

$$\text{Gradient of PQ} = \frac{y_2 - y_1}{x_2 - x_1} = \frac{-8 - (-2)}{5 - 3}$$
$$= \frac{-6}{2} = -3$$

This means,

(i) the line PQ makes an angle θ with the x-axis such that $\tan \theta = -3$,

i.e. $\quad m_{PQ} = \tan \theta = -3$

(ii) that y changes at the rate of -3 units for unit increase in x.

1.3 SECANTS AND TANGENTS

A **secant** is a straight line which cuts a curve in two distinct points. In figure 2, SPQR is a secant to a curve with PQ a chord. P is any point on the curve and Q is a neighbouring point.

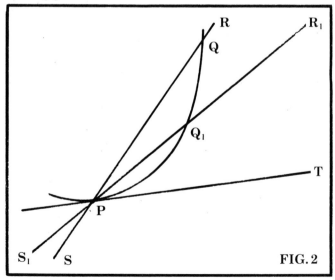

FIG. 2

Keep P fixed and let Q approach P along the curve. The secant SPQR will take up new positions $S_1PQ_1R_1$, $S_2PQ_2R_2$, etc.

As Q approaches P along the curve, the chord PQ approaches nearer and nearer to a certain straight line PT, which passes through P and though produced indefinitely in either direction, does not meet the curve again.

The straight line PT is called the **tangent** to the curve at P.

We say that as Q moves or "tends to" P along the curve, the chord PQ "tends to" the tangent to the curve at P.

We use the symbol → (which is read "tends to") and write

as Q → P along the curve

the chord PQ → the tangent to the curve at P

∴ the gradient of the chord PQ → the gradient of the tangent to the curve at P.

i.e. $m_{PQ} \to m_{\text{tangent to the curve at P}}$.

Again, if in figure 3, P is the point (x, y) and Q the point $(x+h, y+k)$ then as Q → P along the curve

k becomes smaller and smaller, i.e. $k \to 0$

h becomes smaller and smaller, i.e. $h \to 0$

and the gradient of the chord PQ becomes the gradient of the tangent PT to the curve at P.

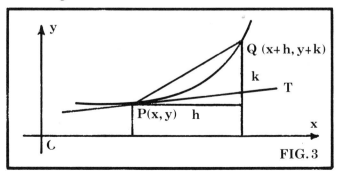

FIG. 3

FINDING THE GRADIENT OF THE TANGENT TO A CURVE 1.4

In figure 4 let $P(x, y)$ and $Q(x+h, y+k)$ be neighbouring points on the curve $y = f(x)$. PQ is the chord through P and Q and PT is the tangent at P.

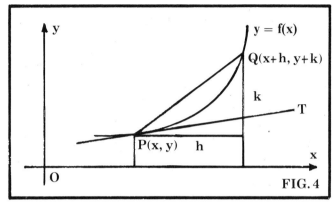

FIG. 4

$$\text{Gradient of PQ} = \frac{y\text{-step}}{x\text{-step}} = \frac{(y+k) - y}{(x+h) - x}$$

$$= \frac{k}{h}$$

When Q → P, $h \to 0$, $k \to 0$, the chord PQ → the tangent PT to the curve at P and $\frac{k}{h} \to$ gradient of the tangent at P.

Example 2

Find the gradient of the tangent to the curve $y = x^2$ at the point

(i) $P(x, y)$

(ii) $A(-2, 4)$.

(i) Let $P(x, y)$ be the point on the curve and $Q(x+h, y+k)$ a neighbouring point on the curve.

Then at P $\quad y = x^2$

and at Q $\quad y+k = (x+h)^2$

Subtracting gives $\quad k = (x+h)^2 - x^2$
$$= x^2 + 2hx + h^2 - x^2$$
$$= 2hx + h^2$$

Hence the gradient of chord $PQ = \dfrac{k}{h} = \dfrac{2hx+h^2}{h}$
$$= 2x+h,$$
$$\text{provided } h \neq 0$$

As $h \to 0$, $\dfrac{k}{h} = 2x+h \to 2x$ and therefore the gradient of the tangent at $P = 2x$.

(ii) At the point $A(-2, 4)$ the gradient of the tangent
$$= 2x \text{ when } x = -2$$
$$= -4.$$

Note

1. $2x$ defines a function called the **gradient** or **derived function** of the function $f: x \to x^2$ which is denoted by $f': x \to 2x$. $2x$ is called the **derivative** of f and the process of finding the derivative of a function is called **differentiation**.

2. If $y = f(x)$, then
   (i) the gradient of the tangent to the curve at P, (i.e. the derivative of f) is written $\dfrac{dy}{dx}$ or $f'(x)$;

   (ii) and $\dfrac{k}{h}$ as $h \to 0$ is written

   "the limit of $\dfrac{k}{h}$" or simply $\lim\limits_{h \to 0} \dfrac{k}{h}$.

Hence in example 2(i) we can rewrite the last sentence as
$$\dfrac{dy}{dx} \text{ or } f'(x) = \lim_{h \to 0} \dfrac{k}{h}$$
$$= \lim_{h \to 0} (2x+h)$$
$$= 2x$$

3. If D is the domain of a given function f then although f' is defined on the whole of D for many functions (f is then said to be **differentiable**), there are some functions for which the domain D_1 (of f') is not equal to D. In such cases D_1 is a subset of D.

Example 3

Find from first principles the derivative of the function f defined by $f(x) = 4x^3$.

Let $y = 4x^3$. Let P be the point (x, y) and $Q(x+h, y+k)$.

Then at P $\quad y = 4x^3$

and at Q $\quad y+k = 4(x+h)^3$

Subtracting gives
$$k = 4(x+h)^3 - 4x^3$$
$$= 4(x^3 + 3x^2h + 3xh^2 + h^3) - 4x^3$$
$$= 12x^2h + 12xh^2 + 4h^2$$

Hence $\dfrac{k}{h} = 12x^2 + 12xh + 4h^2$, provided $h \neq 0$,

and the derivative [i.e. the gradient of the tangent to the curve $y = f(x)$ at $P(x, y)$] is
$$f'(x) \text{ or } \dfrac{dy}{dx} = \lim_{h \to 0} \dfrac{k}{h}$$
$$= \lim_{h \to 0} (12x^2 + 12xh + h^2)$$
$$= 12x^2$$

It follows that the function $f: x \to 4x^3$ has derived function $f': x \to 12x^2$.

Example 4

Find from first principles the derivative of the function f defined by $f(x) = \dfrac{1}{x}$ ($x \neq 0$).

Let $y = \dfrac{1}{x}$ and let P be the point (x, y) and $Q(x+h, y+k)$.

Then at P $\quad y = \dfrac{1}{x}$

and at Q $\quad y+k = \dfrac{1}{x+h}$

Subtracting gives

$$k = \dfrac{1}{x+h} - \dfrac{1}{x} = \dfrac{x-(x+h)}{x(x+h)}$$

$$= \dfrac{-h}{x(x+h)}$$

Hence $\quad \dfrac{k}{h} = \dfrac{-1}{x(x+h)}, \quad h \neq 0$

and $\dfrac{dy}{dx}$ or $f'(x) = \lim\limits_{h \to 0} \dfrac{k}{h} = \lim\limits_{h \to 0} \dfrac{-1}{x(x+h)}$

$$= -\dfrac{1}{x^2}$$

$$= -x^{-2} \quad (x \neq 0)$$

SHORTHAND METHOD OF FINDING DERIVATIVES

Example 5

Find the derivative of $y = f(x)$ at the point $P(x, y)$ where $f(x)$ defines some function f.

Let $P(x, y)$ and $Q(x+h, y+k)$ be neighbouring points.

Then at P $\quad y = f(x)$

and at Q $\quad y+k = f(x+h)$

Subtracting gives

$$k = f(x+h) - f(x)$$

Hence $\quad \dfrac{k}{h} = \dfrac{f(x+h) - f(x)}{h}, \quad h \neq 0$

If $h \to 0$, then

$$\lim\limits_{h \to 0} \dfrac{k}{h} = \lim\limits_{h \to 0} \dfrac{f(x+h) - f(x)}{h}$$

Hence

$$\dfrac{dy}{dx} = f'(x) = \lim\limits_{h \to 0} \dfrac{f(x+h) - f(x)}{h} \quad \ldots\ldots\ldots\ldots (1)$$

This is the derivative of the function f defined by $y = f(x)$ at the point (x, y).

At the point (a, b), the derivative of f is,

$$f'(a) = \lim\limits_{h \to 0} \dfrac{f(a+h) - f(a)}{h} \quad \ldots\ldots\ldots\ldots (2)$$

Example 6

Find the derivative of x^2 from first principles.

Let $f(x) = x^2$ and use formula (1)

Then $f'(x) = \lim\limits_{h \to 0} \dfrac{f(x+h) - f(x)}{h}$ where $f(x) = x^2$

$$= \lim\limits_{h \to 0} \dfrac{(x+h)^2 - x^2}{h}$$

$$= \lim\limits_{h \to 0} \dfrac{x^2 + 2xh + h^2 - x^2}{h}$$

$$= \lim\limits_{h \to 0} (2x + h) = 2x$$

Example 7

Find the derivative of $y = x - \dfrac{1}{x}$, $x \neq 0$ at

(i) the point (x, y), by using formula (1),
(ii) the point $(1, 0)$, by using formula (2) or by substituting in the result of (i).

(i) $f'(x) = \lim\limits_{h \to 0} \dfrac{f(x+h) - f(x)}{h}$

$= \lim\limits_{h \to 0} \dfrac{\left\{(x+h) - \dfrac{1}{x+h}\right\} - \left(x - \dfrac{1}{x}\right)}{h}$

$= \lim\limits_{h \to 0} \dfrac{h - \dfrac{1}{x+h} + \dfrac{1}{x}}{h}$

$= \lim\limits_{h \to 0} \dfrac{\dfrac{hx(x+h) - x + (x+h)}{x(x+h)}}{h}$

$= \lim\limits_{h \to 0} \dfrac{h(x^2 + hx + 1)}{hx(x+h)}$

$= \lim\limits_{h \to 0} \dfrac{x^2 + hx + 1}{x(x+h)}$

$= \dfrac{x^2 + 1}{x^2}$

$= 1 + \dfrac{1}{x^2}$

(ii) (a) $f'(1) = \lim\limits_{h \to 0} \dfrac{f(1+h) - f(1)}{h}$

$= \lim\limits_{h \to 0} \dfrac{\left\{(1+h) - \dfrac{1}{(1+h)}\right\} - \left(1 - \dfrac{1}{1}\right)}{h}$

$= \lim\limits_{h \to 0} \dfrac{\left\{\dfrac{(1+h)^2 - 1}{(1+h)}\right\} - 0}{h}$

$= \lim\limits_{h \to 0} \left\{\dfrac{1 + 2h + h^2 - 1}{h(h+1)}\right\}$

$= \lim\limits_{h \to 0} \dfrac{h(2+h)}{h(h+1)}$

$= \lim\limits_{h \to 0} \dfrac{2+h}{1+h}$

$= 2.$

Or (ii) (b) At the point $(1, 0)$,

$f'(1) = \left[1 + \dfrac{1}{x^2}\right]_{x=1}$

$= 1 + 1$

$= 2.$

1.6 RULES FOR DIFFERENTIATION

From examples 2 to 6 we see that if,

$f(x) = 4x^3$ then $f'(x) = 12x^2$
$f(x) = x^2$ then $f'(x) = 2x$
$f(x) = \dfrac{1}{x} = x^{-1}$ then $f'(x) = -x^{-2}$
$f(x) = x - \dfrac{1}{x}$ then $f'(x) = 1 + x^{-2}$

then we obtain the following rules,

Rule 1. *The derivative of ax^n is nax^{n-1}, $a \in R$, $n \in Q$.*

If $y = -6x^8$, $\dfrac{dy}{dx} = -48x^7$

If $\quad y = \frac{3}{4}x^{-\frac{1}{3}}, \quad \frac{dy}{dx} = -\frac{1}{3} \cdot \frac{3}{4}x^{-\frac{4}{3}} = -\frac{1}{4}x^{-\frac{4}{3}}$

Rule 2. *The derivative of a constant function is the zero function.*

If $\quad y = c \quad \frac{dy}{dx} = 0 \quad c \in R.$

If $\quad y = 8a^2 \quad \frac{dy}{dx} = 0 \quad a \in R.$

Rule 3. *The derivative of the algebraic sum of powers of x is equal to the sum of the derivatives of the terms.*

If $y = x^3 - 2x^2 - 6x + 7$ then $\frac{dy}{dx} = 3x^2 - 4x - 6$

Rule 4. *The derivative of a function of a function—i.e. the chain rule for differentiation.*

If y is a function of u and u is a function of x, then y is also a function of x (via u).

For example, if $y = f(u)$ and $u = g(x)$, then
$$y = f[g(x)] = (f \circ g)(x).$$

Hence a composite function or a *function of a function* is obtained. The rule for differentiation is

$$\frac{dy}{dx} = \frac{df}{du} \cdot \frac{dg}{dx} = \frac{dy}{du} \times \frac{du}{dx}$$

Example 8

Find the derivative of $y = (3 - 4x)^5$.

Let $\quad u = 3 - 4x \quad$ then $\quad y = u^5$

Hence $\quad \frac{dy}{du} = 5u^4 \quad$ and $\quad \frac{du}{dx} = -4$

But $\quad \frac{dy}{dx} = \frac{dy}{du} \times \frac{du}{dx} = 5u^4(-4) = -20(3 - 4x)^4$

Note that this can be written $(5)(-4)(3 - 4x)^{5-1}$. Hence

Rule 5. If $y = (ax + b)^n$ then

$$\frac{dy}{dx} = na(ax + b)^{n-1} \quad a, b \in R, \ n \in Q$$

Example 9

Given $f(x) = \left(x^2 + \frac{1}{x}\right)^2$, $x > 0$, find $f'(x)$.

We can do this in two ways.

Method 1

$$f(x) = \left(x^2 + \frac{1}{x}\right)^2 = x^4 + 2 \cdot x^2 \cdot \frac{1}{x} + \frac{1}{x^2}$$
$$= x^4 + 2x + x^{-2}$$

$$f'(x) = 4x^3 + 2 - 2x^{-3} = 4x^3 + 2 - \frac{2}{x^3}$$

Method 2

$$f(x) = \left(x^2 + \frac{1}{x}\right)^2$$

Let $\quad u = x^2 + \frac{1}{x}$

then $\quad y = u^2 \quad$ where $\quad y = f(x)$

Hence

$\frac{du}{dx} = 2x - x^{-2} \quad$ and $\quad \frac{dy}{du} = 2u$

By the chain rule

$$f'(x) = \frac{dy}{dx} = \frac{dy}{du} \times \frac{du}{dx} = 2u(2x - x^{-2})$$

$$= 2\left(x^2+\frac{1}{x}\right)\left(2x-\frac{1}{x^2}\right)$$

$$= 2\left(2x^3-1+2-\frac{1}{x^3}\right)$$

$$= 4x^3+2-\frac{2}{x^3}$$

Example 10

If $f(x) = \dfrac{3x^2+2x-7}{\sqrt{x}}$, $x > 0$, find $f'(x)$.

To differentiate this we must write $f(x)$ in the form

$$f(x) = \frac{3x^2}{x^{\frac{1}{2}}} + \frac{2x}{x^{\frac{1}{2}}} - \frac{7}{x^{\frac{1}{2}}} = 3x^{\frac{3}{2}} + 2x^{\frac{1}{2}} - 7x^{-\frac{1}{2}}$$

Hence

$$f'(x) = 3 \cdot \tfrac{3}{2}x^{\frac{1}{2}} + 2 \cdot \tfrac{1}{2}x^{-\frac{1}{2}} - 7(-\tfrac{1}{2})x^{-\frac{3}{2}}$$

$$= \tfrac{9}{2}x^{\frac{1}{2}} + x^{-\frac{1}{2}} + \tfrac{7}{2}x^{-\frac{3}{2}}$$

$$= \frac{9x^{\frac{1}{2}}}{2} + \frac{1}{x^{\frac{1}{2}}} + \frac{7}{2x^{\frac{3}{2}}}$$

ASSIGNMENT 1.1

Remember: $\dfrac{dy}{dx}$ or $f'(x)$ is the derivative of f and gives the gradient of the tangent to the curve $y = f(x)$ at the point (x, y) on it.

1. Find the gradient of the lines joining the points,
   (i) $(2, 3), (-1, -2)$
   (ii) $(-3, -7), (-1, 5)$
   (iii) $(2, -5), (0, 5)$
   (iv) $(4, -5), (-2, 7)$
   (v) $(4, 10), (4+a, 10+b)$
   (vi) $(3, -7), (4+a, 9-b)$
   (vii) $(x, y), (x+h, y+k)$.

2. Find from first principles [by using the $P(x, y)$ and $Q(x+h, y+k)$ method] the derivative of
   (i) x^2 (ii) x^3 (iii) $\dfrac{1}{x}$, $x \neq 0$
   (iv) x^2+x (v) $3x^3+2$ (vi) $\dfrac{1}{x^2}$, $x \neq 0$

3. By using the shorthand method
$$f'(x) = \lim_{h \to 0} \frac{f(x+h)-f(x)}{h}$$
   find the derivatives of
   (i) x^2 (ii) $2x^3$ (iii) $-\dfrac{1}{x}$, $x \neq 0$
   (iv) $\dfrac{1}{x^2}$, $x \neq 0$ (v) x^2+x (vi) $x-\dfrac{1}{x}$, $x \neq 0$
   (vii) $4x$ (vii) $x-x^2$ (ix) $\dfrac{1}{x^3}$, $x \neq 0$
   (x) $(x-2)(x+2)$

4. State or write down the derivatives of
   (i) x^8 (ii) $6x^4$
   (iii) 13 (iv) $x^{\frac{1}{2}}$, $x > 0$
   (v) $3x^{\frac{3}{2}}$, $x > 0$ (vi) $5\sqrt{x}$, $x > 0$
   (vii) $6x^{\frac{4}{3}}$, $x > 0$ (viii) $\dfrac{1}{\sqrt{x}}$, $x > 0$
   (ix) $\dfrac{1}{x^2}$, $x \neq 0$ (x) $x^{-\frac{1}{3}}$, $x \neq 0$

5. Write down the derivatives of
   (i) $\dfrac{5}{x^5}$, $x \neq 0$ (ii) $\dfrac{1}{4x^2}$, $x \neq 0$
   (iii) $\sqrt{x}+\dfrac{1}{\sqrt{x}}$, $x > 0$ (iv) x^2+2x+1

(v) $3x^4 - \frac{1}{3}x^3 + x$ (vi) $\frac{1}{2}x^4 - x^3$

(vii) $x^3 - 5x + \frac{1}{x}, x \neq 0$ (viii) $(x+1)^2$

(ix) $\frac{2(x^2-1)}{x}, x \neq 0$ (x) $\frac{x^2 - 2x}{\sqrt{x}}, x > 0$

(xi) $\left(x - \frac{1}{x}\right)^2, x \neq 0$.

6. (a) Given $f(x) = \frac{x^3 - 7x^2 + 9}{x}, x > 0$, find $f'(x)$.

(b) Given $f(x) = \frac{2x^2 - 7x + 3}{\sqrt{x}}, x > 0$, find $f'(x)$.

(c) Given $f(x) = \frac{1}{x} - \frac{1}{x^2}, x \neq 0$, find $f'(-2)$.

(d) Given $f(x) = 2x^2 + \frac{2}{x^2}, x \neq 0$, find $f'(-1)$ and $f'(-2)$.

7. Remembering $f(x) = (ax+b)^n$ then $f'(x) = an(ax+b)^{n-1}$, find the derivative of:

(i) $(3x+7)^2$ (ii) $(7-3x)^2$
(iii) $(5x-4)^5$ (iv) $(4+\frac{1}{2}x)^6$
(v) $2(5-2x)^3$ (vi) $\frac{2}{x-1}, x \neq 1$
(vii) $\frac{3}{(5x+2)^3}, x \neq -\frac{2}{5}$ (viii) $(px+q)^n$
(ix) $\sqrt{(4-3x)}, x < \frac{4}{3}$ (x) $(a-bx)^p$

8. Show that $f(x) = x\left(x - \frac{1}{x}\right)\left(x + \frac{1}{x}\right) = x^3 - \frac{1}{x}, x > 0$ and hence find $f'(x)$.

9. Given $f(x) = \left(\sqrt{x} - \frac{1}{\sqrt{x}}\right)^2, x > 0$, find $f'(x)$.

10. Show that for $x > 0$,
$$\left(x + 1 - \frac{1}{x}\right)\left(x - 1 - \frac{1}{x}\right) = x^2 - 3 + \frac{1}{x^2}.$$
If $f(x) = \left(x + 1 - \frac{1}{x}\right)\left(x - 1 - \frac{1}{x}\right), x > 0$, find $f'(x)$.

11. Differentiate with respect to x

(i) $\frac{1}{\sqrt{(3-2x)}}, x \neq \frac{3}{2}$ (ii) $\frac{3x+x^2}{\sqrt{x}}, x > 0$

(iii) $(4x-1)^{\frac{3}{2}}, x > \frac{1}{4}$

12. Use the chain rule to find the derivatives of

(i) $(x^2+1)^3$ (ii) $(4-x^2)^{\frac{1}{2}}, -2 < x < 2$
(iii) $(2x^2+1)^5$ (iv) $\frac{1}{\sqrt{(4-3x)}}, x < \frac{4}{3}$
(v) $\frac{1}{\sqrt{(2x+3)}}, x > -\frac{3}{2}$ (vi) $\frac{1}{x^2-4}, x \neq \pm 2$

APPLICATION OF DERIVATIVES TO TANGENTS   1.7

The straight line of gradient m and passing through the point $P(a, b)$ has equation $(y - b) = m(x - a)$.

Hence to find the equation of a straight line we require to know two facts, namely,

(i) the coordinates of two points through which the line passes; or
(ii) the gradient of the line and the coordinates of one point through which the line passes.

Example 11

Find the equation of the tangent to the curve

$$y = \frac{2+x}{x}, \quad (x \neq 0)$$

at the point where $x = -1$.
  When $x = -1$,

$$y = \frac{2-1}{-1} = -1, \quad \text{i.e. the point is P}(-1, -1).$$

Now the gradient of the tangent at $P(-1, -1)$ is the value of $f'(x)$, i.e. of $\frac{dy}{dx}$ at the point where $x = -1$.

But $$y = \frac{2+x}{x} = \frac{2}{x} + \frac{x}{x}$$
$$= 2x^{-1} + 1$$

Hence $$\frac{dy}{dx} = -2x^{-2} = -\frac{2}{x^2}$$

At $x = -1$, $\frac{dy}{dx} = -\frac{2}{(-1)^2} = -2.$

Hence the gradient of the tangent to the curve at $P(-1, -1)$ is $m = -2$ and the equation of the tangent is $(y-b) = m(x-a)$, where $m = -2$ and (a,b) is the point $P(-1, -1)$.
Therefore tangent has equation

$$(y+1) = -2(x+1)$$
$$\Rightarrow \quad y+1 = -2x-2$$
$$\Rightarrow y+2x+3 = 0$$

Example 12

Find the equation of the tangent to the curve $y = x^3 - 4x$ at the point where $x = 1$, and show that this tangent meets the curve again on the x-axis. Find the coordinates of the point on the curve at which a parallel tangent can be drawn.
  When $x = 1$, $y = 1 - 4 = -3$, i.e. the point is $P(1, -3)$.
  Gradient of the tangent at $P(1, -3)$ is the value of $f'(x)$, i.e. of $\frac{dy}{dx}$ at the point where $x = 1$.

But $y = x^3 - 4x$, hence $\frac{dy}{dx} = 3x^2 - 4.$

At $x = 1$, $\frac{dy}{dx} = 3 - 4 = -1.$

Hence the gradient of the tangent to the curve at $P(1, -3)$ is $m = -1$ and the equation of the tangent is $(y-b) = m(x-a)$, where $m = -1$ and (a,b) is the point $P(1, -3)$.

$$\Rightarrow y+3 = -1(x-1)$$
$$\Rightarrow y+x+2 = 0$$

The tangent $y + x + 2 = 0$ meets the curve $y = x^3 - 4x$ at the points whose x coordinates are given by,

$$x^3 - 4x = -x - 2$$
$$\Leftrightarrow x^3 - 3x + 2 = 0$$

Let $f(x) = x^3 - 3x + 2$ and since the line $y + x + 2 = 0$ is a tangent to the curve $y = x^3 - 4x$ at the point where $x = 1$, then $x = 1$ is a solution of $f(x) = 0 \Rightarrow (x-1)$ is a factor of $f(x)$.
  Check that $x = 1$ is a solution of $f(x) = 0$.

i.e. $\qquad f(1) = 1^3 - 3.1 + 2 = 1 - 3 + 2 = 0$
$$\Rightarrow (x-1) \text{ is a factor of } f(x)$$

Hence $\quad f(x) = x^3 - 3x + 2 = (x-1)(x^2 + x - 2)$
$$= (x-1)(x-1)(x+2)$$

and $f(x) = 0 \Leftrightarrow (x-1)(x-1)(x+2) = 0$
$\Leftrightarrow x = 1$ (twice) or $x = -2$

i.e. the tangent meets the curve again at the point where $x = -2$.
When $x = -2$,
$$y = x^3 - 4x$$
$$\Rightarrow y = (-2)^3 - 4(-2)$$
$$= -8 + 8$$
$$= 0$$

i.e. the tangent meets the curve again at the point $(-2, 0)$.
i.e. on the x-axis.

The tangent $y + x + 2 = 0$ to the curve $y = x^3 - 4x$ has gradient -1.

The tangent at any point $P(x, y)$ on the curve has gradient $\dfrac{dy}{dx} = 3x^2 - 4$.

Hence the coordinates of the point on the curve at which a parallel tangent can be drawn is given by
$$3x^2 - 4 = -1$$
$$\Rightarrow 3x^2 = 3$$
$$\Rightarrow x^2 = 1$$
$$\Rightarrow x = \pm 1$$

i.e. a parallel tangent can be drawn at the point on the curve where $x = -1$.
When $x = -1$,
$$y = x^3 - 4x$$
$$\Rightarrow y = -1 + 4$$
$$= 3$$

i.e. at the point $(-1, 3)$.

ASSIGNMENT 1.2

1. Find the gradient of the tangent to the following curves at the given points [i.e. find the value of $\dfrac{dy}{dx}$ or $f'(x)$ at the given points].
   (i) $y = x^2$ at $(-2, 4)$
   (ii) $y = 3x^2 + 9$ at $(1, 12)$
   (iii) $f(x) = (1-2x)(1+x)$ at $(0, 1)$
   (iv) $f(x) = 4x^3 - 3x^2 + 6x + 1$ at the point where $x = -1$
   (v) $y = 3x^3 - x$ at (x, y)
   (vi) $y = \dfrac{1}{x}$ at the point (a, b), $a \neq 0$

2. For each of the following curves find the gradient of the tangent and the equation of the tangent at the given point.
   (i) $y = x^2 - 2$ at $(2, 2)$
   (ii) $y = x^3$ at $(-2, -8)$
   (iii) $y = 2x - 1$ at $x = -1$
   (iv) $y = (x-1)^2$ at $x = 1$
   (v) $y = 3$ at $x = k$
   (vi) $y = x^3 - 2x^2 - 6$ at $(3, 3)$

3. Find the equation of the tangent to the curve $y = x^3 - 2x^2 + 1$ at the point where $x = 2$.

4. Find the equation of the tangent to the curve $yx = 1 - x$ at the point where $x = -1$.

5. Find the equation of the tangent to the curve $y = x(x-1)$ at the point given by $x = 3$. If this tangent cuts the coordinate axes at A and B find the coordinates of A and B.

6. Show that the gradient of the tangent to the curve $y = x + \dfrac{1}{x}$ at the point where $x = 2$ has value $\tfrac{3}{4}$.

   Find the coordinates of the other point on the curve at which the tangent has gradient $\tfrac{3}{4}$.

7. Find the coordinates of the points on the curve $y = x^3 + 3x^2 - 9x + 4$ at which the tangents are parallel to the x-axis.

8. Find the equation of the tangent to the curve $y = x(x^2 - 1)$ at the point given by $x = 1$. Find the coordinates of the point at which this tangent meets the curve again.

9. Show that at all points on the curve $y = x^3 + 2x - 1$ the gradient of the tangent is positive. Find the equation of the tangent at the point $P(-1, -4)$ on the curve and determine the coordinates of the point Q in which the tangent cuts the curve again.

1. If the function f defined by $f(x) = x^2 - 2$ is differentiated from first principles, then $f'(x)$ is equal to

   A. $\lim\limits_{h \to 0} \dfrac{2xh + h^2}{h}$

   B. $\lim\limits_{h \to 0} (2xh + h^2)$

   C. $\lim\limits_{h \to 0} \dfrac{(x+h)^2 - 2}{h}$

   D. $\lim\limits_{h \to 0} \dfrac{2xh + h^2 - 2}{h}$

   E. $\lim\limits_{h \to 0} \dfrac{h^2 - 2}{h}$

2. The function f has a derivative defined by
$$f'(x) = \lim_{h \to 0} \dfrac{-h}{(x+a)(x+a+h)h}$$
   Which one of the following statements is true?

   A. $f'(x) = -\dfrac{1}{x^2}$

   B. $f'(a) = -\dfrac{1}{4a^2}, \quad a \neq 0$

   C. $f'(0) = -\dfrac{1}{a(a+h)}$

   D. $f'(0) = -\dfrac{1}{x^2}$

   E. $f'(a)$ has no value, $a \in R$

3. $f(x) = \dfrac{1}{x}, x \neq 0, f'(-2)$ equals

   A. -1 D. $\tfrac{1}{4}$

   B. $-\tfrac{1}{2}$ E. $\tfrac{1}{2}$

   C. $-\tfrac{1}{4}$

ASSIGNMENT 1.3

Objective type items testing Sections 1.1–1.7.
Instructions for answering these items are given on page 16.

4. $f(x) = (1-3x)^3$. $f'(x)$ equals
   A. $-9(1-3x)^2$
   B. $9(1-3x)^2$
   C. $3(1-3x)^2$
   D. $-3(1-3x)^4$
   E. $3(1-3x)^4$

5. $\lim_{h \to 0} \dfrac{\left\{\dfrac{1}{x+h} - \dfrac{1}{x}\right\}}{h}$ equals

   A. $-\dfrac{1}{x+h}$
   B. $-\dfrac{1}{x^2}$
   C. 0
   D. $\dfrac{1}{x^2}$
   E. none of these

6. $f(x) = \dfrac{1}{4+3x}$, $(x \neq -\frac{4}{3})$. $f'(x)$ is equal to

   A. $\dfrac{-3}{(4+3x)^2}$
   B. $\dfrac{-1}{(4+3x)^2}$
   C. $\dfrac{-1}{3(4+3x)^2}$
   D. $\dfrac{1}{(4+3x)^2}$
   E. $\dfrac{3}{(4+3x)^2}$

7. The gradient of the tangent to the curve $y = x^2 - 1$ at the point $(0, -1)$ is
   A. 0
   B. 1
   C. -1
   D. 2
   E. has no finite value

8. The gradient of the tangent to the curve $y = \frac{1}{4}x^2 - x^{-1}$ at the point where $x = -1$ is
   A. $-1\frac{1}{2}$
   B. $-\frac{1}{2}$
   C. 0
   D. $\frac{1}{2}$
   E. $1\frac{1}{2}$

9. If $f(x) = \dfrac{1}{x+k}$ then $f'(-2) = -4$ if k has the value
   (1) $1\frac{1}{2}$
   (2) $3\frac{1}{2}$
   (3) 0

10. If $f(x) = \dfrac{1}{x}$, which of the following is/are false?
    (1) $f'(x)$ does not exist at $x = 0$
    (2) $f'(1) = -1$
    (3) $f'(2x) = 2f'(x)$

11. (1) $f(x) = \dfrac{1}{x-a}$
    (2) $f(x)$ is not differentiable at $x = a$

12. (1) $f(x) = x + k$, $k \in R$
    (2) $f'(x) = 1$

137

UNIT 2: STATIONARY, TURNING AND INFLEXION POINTS

INCREASING AND DECREASING FUNCTIONS 2.1
STATIONARY POINTS AND STATIONARY VALUES

Consider the function f defined by $f(x) = 1 - x^2$. The graph of the function is shown in figure 5.

If $f(x) = 1 - x^2$ then $f'(x) = -2x$.

(i) If $-1 < x < 0$ then $f'(x) > 0$,

$$\left[\text{i.e. } f'(x) \left(\text{or } \frac{dy}{dx} \right) \text{ is positive} \right].$$

Hence as x increases from -1 to 0, $f'(x) > 0$ and the value of $f(x)$ increases.

i.e. between -1 and 0 f is said to be increasing.

(ii) If $0 < x < 1$ then $f'(x) < 0$,

$$\left[\text{i.e. } f'(x) \left(\text{or } \frac{dy}{dx} \right) \text{ is negative} \right].$$

Hence as x increases from 0 to 1, $f'(x) < 0$ and the value of $f(x)$ decreases.

i.e. between 0 and 1 f is said to be decreasing.

Hence Note

(i) For f to be increasing $f'(x) \left(\text{or } \frac{dy}{dx} \right)$ must be > 0.

(ii) For f to be decreasing $f'(x) \left(\text{or } \frac{dy}{dx} \right)$ must be < 0.

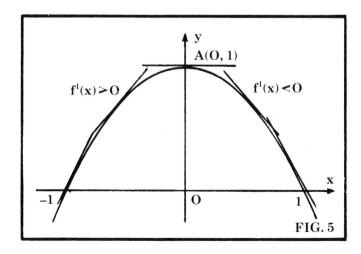

FIG. 5

The point A on the curve with coordinates $(0, 1)$ and where $f'(x) = 0$ is called a **Stationary Point** because at A f is neither increasing nor decreasing.

Hence points on a curve where $f'(x)$ $\left(\text{or } \dfrac{dy}{dx}\right) = 0$ are called **Stationary Points.**

The values of $f(x)$ [i.e. the values of the y-coordinates] at such points are called the **Stationary Values** of the function f.

Example 1

For what values of x is (i) $x - 1 > 0$
                              (ii) $x + 1 < 0$
                              (iii) $(x-1)(x+1) < 0$
                              (iv) $(x-1)(x+1) > 0$?

(i) $x - 1 > 0 \Leftrightarrow x > 1$

(ii) $x + 1 < 0 \Leftrightarrow x < -1$

(iii) $(x-1)(x+1) < 0 \Leftrightarrow x < 1 \text{ or } x > -1$

Test this for several values of $x < 1$ and $x > -1$. This range of values of x may be written

$(x-1)(x+1) < 0 \Leftrightarrow -1 < x < 1$

(iv) $(x-1)(x+1) > 0 \Leftrightarrow x > 1 \text{ or } x < -1$

Test this for several values of $x > 1$ and $x < -1$.

Example 2

Find the intervals for which the function f defined by $f(x) = 4 + 2x^2 - \tfrac{1}{3}x^3$ is (i) increasing, (ii) decreasing.

Here $f'(x) = 4x - x^2$

(i) If f is increasing $f'(x) > 0$

$$f'(x) > 0 \Leftrightarrow 4x - x^2 > 0$$
$$\Leftrightarrow x(4 - x) > 0$$

But $x(4 - x) > 0 \Leftrightarrow 0 < x < 4$

It follows that f is increasing in the interval for which $0 < x < 4$.

(ii) If f is decreasing $f'(x) < 0$

$$f'(x) < 0 \Leftrightarrow 4x - x^2 < 0$$
$$\Leftrightarrow x(4 - x) < 0$$

But $x(4-x) < 0 \Leftrightarrow x < 0 \text{ or } x > 4$

Hence f is decreasing in the interval for which $x < 0$ and in the interval for which $x > 4$.

Example 3

Find the stationary values of the function f defined by $f(x) = \tfrac{1}{3}x^3 - x$ and the stationary points on the graph of the function.

If $f(x) = \tfrac{1}{3}x^3 - x$ then $f'(x) = x^2 - 1$.

For stationary values

$$f'(x) = 0 \Leftrightarrow x^2 - 1 = 0$$
$$\Leftrightarrow x^2 = 1$$
$$\Leftrightarrow x = 1 \text{ or } -1$$

Now $f(1) = \tfrac{1}{3} - 1 = -\tfrac{2}{3}$
and $f(-1) = -\tfrac{1}{3} + 1 = \tfrac{2}{3}$

Hence the stationary values are $-\tfrac{2}{3}$ and $\tfrac{2}{3}$, and the stationary points are $(1, -\tfrac{2}{3})$ and $(-1, \tfrac{2}{3})$.

ASSIGNMENT 2.1

1. For what values of x is
    (i) $x - 2 > 0$
    (ii) $x + 2 < 0$
    (iii) $(x-2)(x+2) > 0$
    (iv) $(x-2)(x+2) < 0$?

2. For what values of x is
   (i) $x - 3 < 0$
   (ii) $x + 3 > 0$
   (iii) $(x-3)(x+3) > 0$
   (iv) $(x-3)(x+3) < 0$?

3. $f(x) = x(x-1) > 0$, state the range of values of x.

4. $f(x) = x(2x+1) < 0$, state the range of values of x.

5. For what range of values of x is $-1 \leq 2x - 3 \leq 1$?

6. If $f(x) = (x+2)(2x-3) > 0$, find the range of values of x.

In questions 7 to 16 determine the intervals in which each function defined by $f(x)$ is increasing and the intervals in which it is decreasing.

7. $f(x) = x^3$
8. $f(x) = 2x^2$
9. $f(x) = x^2 - 2x$
10. $f(x) = 2x - x^2$
11. $f(x) = x^3 - 12x$
12. $f(x) = x^2 - 6x + 4$
13. $f(x) = \frac{1}{3}x^3 - x^2$
14. $f(x) = x(2-x)^2$
15. $f(x) = 2x^2(x-3)$
16. $f(x) = x - \frac{1}{x}$, $x \neq 0$

17. Show that for all values of $x \in Q$, the function f defined by $f(x) = 2(1-3x)^3$ is a decreasing function.

18. For what values of $a \in Q$ is the function f defined by $f(x) = (ax+b)^3$
    (i) increasing,
    (ii) decreasing.

19. If $f(x) = 2x^3 - x^4$, find whether $f(x)$ is increasing, decreasing or stationary when,
    (i) $x = 1\frac{1}{2}$
    (ii) $x = 3$
    (iii) $x = -1$

20. Find the stationary values and the stationary points on the graph of the function f defined by
$$f(x) = x^4 - 4x^3 + 4x^2 + 2.$$

TURNING POINTS AND TURNING VALUES 2.2

(i) **The Maximum Turning Point**

In figure 6 let the point where $x = a$ on the curve $y = f(x)$ be a stationary point.

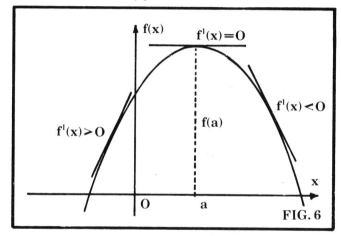

FIG. 6

In this case we have that when

$$x < a, \quad f'(x) > 0$$
$$x = a, \quad f'(x) = 0$$
$$x > a, \quad f'(x) < 0$$

Hence if $f'(x)$ $\left(\text{i.e. } \frac{dy}{dx}\right)$ changes sign from positive to negative through some point $[a, f(a)]$, the stationary point is called a **Maximum Turning Point**. The corresponding value of $f(x)$ [i.e. $f(a)$] is called the **Maximum Turning Value** of the function f.

(ii) The Minimum Turning Point

In figure 7 let the point where $x = b$ on the curve $y = f(x)$ be a stationary point.

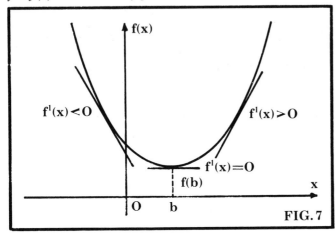

FIG. 7

In this case we have that when
$$x < b, \quad f'(x) < 0$$
$$x = b, \quad f'(x) = 0$$
$$x > b, \quad f'(x) > 0$$

Hence if $f'(x)$ (i.e. $\frac{dy}{dx}$) changes sign from negative to positive through some point $[b, f(b)]$, the stationary point is called a **Minimum Turning Point**. The corresponding value of $f(x)$ [i.e. $f(b)$] is called the **Minimum Turning Value** of the function f.

Example 4

Find the stationary values of the function f defined by $f(x) = 2x^3 - 3x^2 - 12x + 1$ and determine their nature.

(*Note:* Determine their nature means determine which stationary values are maxima and which are minima.)

$$f(x) = 2x^3 - 3x^2 - 12x + 1$$
$$f'(x) = 6x^2 - 6x - 12 = 6(x^2 - x - 2)$$
$$= 6(x+1)(x-2)$$

Stationary values are given by
$$f'(x) = 0 \Leftrightarrow 6(x+1)(x-2) = 0$$
$$\Leftrightarrow x = -1 \text{ or } 2$$

Hence the stationary values are
$$f(-1) = -2 - 3 + 12 + 1 = 8$$
and $$f(2) = 16 - 12 - 24 + 1 = -19$$

and the stationary points are $(-1, 8)$ and $(2, -19)$.

To determine the nature of the stationary values (or points) we examine the sign of $f'(x)$ for values of x slightly less and slightly greater than -1 and values of x slightly less and slightly greater than 2.

To do this we draw up a table of signs for
$$f'(x) = 6(x+1)(x-2)$$

If x is slightly < -1, $f'(x) > 0$ and the tangent to the graph of $f(x)$ slopes thus ↗.
If x is slightly > -1, $f'(x) < 0$ and the tangent to the graph of $f(x)$ slopes thus ↘.
If x is slightly > 2, $f'(x) > 0$ and the tangent to the graph of $f(x)$ slopes thus ↗.
At $x = -1$ and $x = 2$, $f'(x) = 0$ and the tangent to the graph of $f(x)$ is parallel to the x-axis thus →.

Hence the table

x	→	-1	→	2	→
$f'(x)$	+	0	−	0	+
$f(x)$	↗	Max. →	↘	Min. →	↗

Hence

$f(-1) = 8$ is a maximum turning value,
$f(2) = -19$ is a minimum turning value, and
$(-1, 8)$ is a maximum turning point,
$(2, -19)$ is a minimum turning point.

POINTS OF INFLEXION

Consider the function f defined by

$$f(x) = 3x^5 - 5x^3 + 1$$
$$f'(x) = 15x^4 - 15x^2$$
$$= 15x^2(x^2 - 1) = 15x^2(x-1)(x+1)$$

For stationary values

$$f'(x) = 0 \Leftrightarrow 15x^2(x-1)(x+1) = 0$$
$$\Rightarrow x = 0, 1 \text{ or } -1$$

For the nature of the stationary values we use the table

x	$\to$	-1	$\to$	0	$\to$	1	$\to$
$f'(x)$	$+$	0	$-$	0	$-$	0	$+$
$f(x)$	$\nearrow$	Max.	$\searrow$	$\to$ P.I. $\searrow$	$\to$	Min. $\to$	$\nearrow$

From the table the points given by $x = -1$ and $x = 1$ give maximum and minimum turning values with corresponding maximum and minimum turning points.
Although $f'(x) = 0$ when $x = 0$, $f'(x)$ does not change sign as x passes through the value $x = 0$ even though the tangent to the graph of the function is parallel to the x-axis. Such a point is called a **Point of Inflexion**.
When $x = 0$, $f(0) = 1$ and the point of inflexion is at $(0, 1)$.
On either side of the point $(0, 1)$ the gradient of the curve is negative, so that f decreases through this stationary point and therefore the curve crosses its tangent at this point $(0, 1)$. This is shown in figure 8.

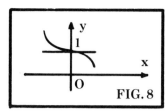
FIG. 8

Note: At a point of inflexion the curve *crosses its tangent* and such points can occur when the tangent is not parallel to the x-axis.

The four sketches (figures 9–12) show various points of inflexion.

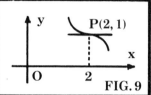
FIG. 9

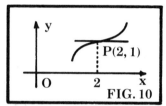
FIG. 10

Here $\dfrac{dy}{dx}$ or $f'(x) = 0$ at $x = 2$. y decreases as x increases and $\dfrac{dy}{dx}$ remains negative through $x = 2$. This is a point of inflexion on a **falling curve**.

Here $\dfrac{dy}{dx}$ or $f'(x) = 0$ at $x = 2$. y increases as x increases and $\dfrac{dy}{dx}$ remains positive through $x = 2$. This is a point of inflexion on a **rising curve**.

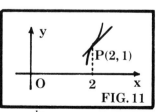
FIG. 11

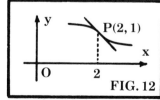
FIG. 12

Here $\dfrac{dy}{dx} > 0$ at the point of inflexion.

Here $\dfrac{dy}{dx} < 0$ at the point of inflexion.

Example 5

Find the stationary values of the function f defined by $f(x) = 2x^3 - 3x^4$, the stationary points on the graph of the function and the nature of the stationary points.

$$f(x) = 2x^3 - 3x^4$$
$$f'(x) = 6x^2 - 12x^3 = 6x^2(1-2x)$$

For stationary values (or points)

$$f'(x) = 0 \Leftrightarrow 6x^2(1-2x) = 0$$
$$\Rightarrow x = 0 \quad \text{or} \quad x = \tfrac{1}{2}$$

Hence the stationary values are $f(0) = 0$ and

$$f(\tfrac{1}{2}) = 2 \times \tfrac{1}{8} - 3 \times \tfrac{1}{16} = \tfrac{1}{16}$$

and the stationary points are $(0,0)$ and $(\tfrac{1}{2}, \tfrac{1}{16})$.

For the nature of the stationary values (or points) we use the table.

x	$\rightarrow$	0	$\rightarrow$	$\tfrac{1}{2}$	$\rightarrow$
$f'(x)$	+	0	+	0	−
$f(x)$	↗	P.I.	↗	Max.	↘

At $x = 0$ the stationary point is a point of inflexion.
At $x = \tfrac{1}{2}$ the stationary point is a maximum turning point, and $f(\tfrac{1}{2}) = \tfrac{1}{16}$ is the maximum turning value.

ASSIGNMENT 2.2

In questions 1 to 10, find the stationary values and stationary points on the graphs of the functions defined by $f(x)$ or y and determine their natures.

1. $f(x) = x^2 - 2x$
2. $f(x) = 2x^3 + 3x^2 - 12x + 1$
3. $f(x) = 3x - x^3$
4. $f(x) = 2 - 3x - 2x^2$
5. $f(x) = x(x^2 - 9)$
6. $f(x) = 3x^3$
7. $f(x) = x^3(x+4)$
8. $f(x) = x(x-3)^2$
9. $y = x^4 - 2x^2$
10. $y = 3x^4 - 2x^3$

In questions 11 to 16, find the maximum and minimum values of the functions defined by

11. $f(x) = 3x - x^2 - \dfrac{x^3}{3}$
12. $f(x) = 3x^2 - 6x + 1$
13. $f(x) = 8 - x^2$
14. $f(x) = 2x^3 - 15x^2 + 36x$
15. $y = 6 + 12x - 3x^2 - 2x^3$
16. $y = x(x+1)^2$

17. Show that the function defined by $f(x) = qx^2 - px + 4$, $p, q \neq 0$ has a stationary value at $x = \tfrac{1}{4}$ if and only if $q = 2p$. If $q = 2p$ determine the nature of this stationary value when $p > 0$.

ASSIGNMENT 2.3

Objective type items testing Sections 2.1–2.3.
Instructions for answering these items are given on page 16.

1. For the function f defined by

$$f(x) = x + \frac{1}{x}$$

A. $f(x)$ decreases for $x < -1$ and $x > 1$
B. $f(x)$ decreases for $-1 < x < 0$ and $0 < x < 1$
C. $f(x)$ increases for $-1 < x < 1$
D. $f(x)$ decreases for $-1 < x < 0$ but increases for $0 < x < 1$
E. $f(x)$ increases only if $x > 1$

2. The function f defined by $f(x) = x^3 - 3ax$ increases for all values of x,
   A. if $a < 0$
   B. if $a = 0$
   C. if $a > 0$
   D. and for all values of $a \in R$
   E. but never increases for $a \in R$

3. $f(x) = (px - 1)^2$. Therefore,
   A. $f(x)$ has a maximum stationary value if $p > 0$
   B. $f(x)$ has a maximum stationary value if $p < 0$
   C. $f(x)$ has a minimum stationary value for all values of p except $p = 0$
   D. $f(x)$ has a stationary value only if $p = 0$
   E. $f(x)$ has no stationary values

4. The function f defined by $f(x) = (x - 1)^3$ has
   A. a point of inflection only
   B. one maximum and one minimum turning point only
   C. no stationary points
   D. one maximum turning point, one minimum turning point and one point of inflexion
   E. only one turning point

5. $f(x) = (x + 1)(x - 3)$ has a stationary value when x equals
   A. -3
   B. -1
   C. 0
   D. 1
   E. 3

6. The function defined by $f(x) = 3x(x - 6)$ is increasing when the value of x is
   A. -6
   B. -3
   C. 0
   D. 3
   E. 6

7. The description of the turning value of the function f defined by $f(x) = -2x^2 - 1$ is
   A. Minimum; -3
   B. Maximum; -2
   C. Minimum; -1
   D. Maximum; -1
   E. Maximum; 0

8. If $\dfrac{dy}{dx} > 0$ for all values of x then
   A. $y > 0$ for all x
   B. y cannot have a maximum value
   C. y is decreasing
   D. $y < 0$ for all x
   E. none of these

9. $f(x) = x^4 - 3x^3$ then
   (1) $f(x)$ is decreasing when $x = 2$
   (2) $f(x)$ is decreasing when $x = -2$
   (3) $f(x)$ is increasing when $x = 0$

10. $f(x) = x^2 - 2x^4$. For which of the following value(s) of x has the function a stationary value?
   - (1) $-\frac{1}{2}$
   - (2) 0
   - (3) $\frac{1}{2}$

11. (1) A curve $y = f(x)$ has a minimum turning point at $(2, 0)$.
    (2) $f(x) = \frac{1}{2}x^2 - 2x$.

12. (1) The curve $y = x^3 - 1$ has only one stationary point.
    (2) The stationary point is a point of inflexion.

UNIT 3: GRAPHS

PRACTICAL APPLICATION OF STATIONARY VALUES

We now use our knowledge of stationary values and stationary points to sketch the graphs of certain functions and to solve certain types of problems involving maximum and minimum turning values.

SKETCHING THE GRAPHS OF CERTAIN FUNCTIONS

Our ability to sketch quickly the graphs of functions is useful in many branches of mathematics especially in further work on the calculus.

In curve sketching we make use of the following:

(i) We find the points where the graph cuts the coordinate axes, i.e.
  (a) Solve the equation $f(x) = 0$ or $y = 0$ to obtain the point or points of intersection with the x-axis.
  (b) Find the value of $f(0)$ to obtain the point of intersection with the y-axis.

(ii) We require the **behaviour** of $f(x)$ for large positive and large negative values of x. The behaviour usually depends on the highest power of x in $f(x)$.

If $f(x) = -x^3$, then if x is large and positive, $f(x)$ is large and negative. But if x is large and negative, then $f(x)$ is large and positive.

This information is not required if the graph is to be drawn within a certain interval.

(iii) We look for the stationary points and stationary values.

(iv) We want the nature of the stationary points.

Example 1

Sketch the graph of the function f defined by

$$f(x) = x^3 - 3x$$

(i) **Intersection with the axes**

$f(0) = 0$; hence when $x = 0$, $y = 0$, i.e. the point $(0, 0)$ lies on the graph.

$$f(x) = 0 \Leftrightarrow x^3 - 3x = 0$$
$$\Leftrightarrow x(x^2 - 3) = 0$$
$$\Leftrightarrow x = 0 \text{ or } x^2 = 3$$
$$\Rightarrow x = 0 \text{ or } \sqrt{3} \text{ or } -\sqrt{3}$$

Hence the points $(0, 0)$, $(\sqrt{3}, 0)$, $(-\sqrt{3}, 0)$ lie on the graph of the function.

(ii) **Behaviour of $f(x)$ for large x**

$f(x) = x^3 - 3x$ and the behaviour depends on x^3.

If x is large and negative, x^3 is large and negative, i.e. $f(x)$ is large and negative.

Hence the graph starts in the 3rd quadrant.

If x is large and positive, x^3 is large and positive, i.e. $f(x)$ is large and positive.

Hence the graph finishes in the 1st quadrant.

(iii) **Stationary points**

$$f(x) = x^3 - 3x$$
$$f'(x) = 3x^2 - 3$$
and $$f'(x) = 0 \Leftrightarrow 3(x^2 - 1) = 0$$
$$\Leftrightarrow x^2 = 1$$
$$\Rightarrow x = 1 \text{ or } -1$$

Hence $f(1) = -2$, $f(-1) = 2$ are the stationary values and the stationary points are $(1, -2)$ and $(-1, 2)$.

3.1

3.2

(iv) **Nature of the stationary points**—use the table.

x	$\to$	-1	$\to$	1	$\to$
$f'(x)$	$+$	0	$-$	0	$+$
$f(x)$	$\nearrow$	Max.	$\searrow$	$\to$ Min.	$\nearrow$

Hence $(1, -2)$ is a minimum turning point and $(-1, 2)$ is a maximum turning point.

(v) **Sketch**

To draw the sketch we have the following information:

The graph starts in the 3rd quadrant; passes through the point $(-\sqrt{3}, 0)$; has a maximum turning point at $(-1, 2)$; passes through the point $(0, 0)$; has a minimum turning point at $(1, -2)$; passes through the point $(\sqrt{3}, 0)$ and finishes in the 1st quadrant.

Hence sketch (figure 13),

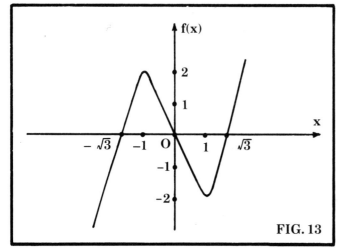

FIG. 13

Example 2

Find the stationary values and stationary points on the graph of the function f defined by $f(x) = x^2 - x^3$ and determine their nature. Sketch the graph of the function.

Stationary values

$$f(x) = x^2 - x^3,$$
$$f'(x) = 2x - 3x^2$$

and $\quad f'(x) = 0 \Leftrightarrow 2x - 3x^2 = 0$

$$\Leftrightarrow x(2 - 3x) = 0$$
$$\Rightarrow \quad x = 0 \text{ or } \tfrac{2}{3}$$

$f(0) = 0$ and $f(\tfrac{2}{3}) = \tfrac{4}{9} - \tfrac{8}{27} = \tfrac{4}{27}$ are the stationary values and the stationary points are $(0, 0)$ and $(\tfrac{2}{3}, \tfrac{4}{27})$.

Nature of the stationary points

x	$\to$	0	$\to$	$\tfrac{2}{3}$	$\to$
$f'(x)$	$-$	0	$+$	0	$-$
$f(x)$	$\searrow$	$\to$ Min.	$\nearrow$	$\to$ Max.	$\searrow$

Hence $(0, 0)$ is a minimum turning point and $(\tfrac{2}{3}, \tfrac{4}{27})$ is a maximum turning point.

Intersection with the axes

$f(0) = 0$; hence when $x = 0$, $y = 0$

$$f(x) = 0 \Leftrightarrow x^2 - x^3 = 0$$
$$\Leftrightarrow x^2(1 - x) = 0 \Rightarrow x = 0 \text{ (twice) or } x = 1$$

i.e. the points $(0, 0)$, $(1, 0)$ are on the graph.

Behaviour of $f(x)$ for large x

$f(x) = x^2 - x^3$ and the behaviour depends on $-x^3$.

If x is large and negative, $-x^3$ is large and positive, i.e. $f(x)$ is large and positive.

Hence the graph starts in the 2nd quadrant.

If x is large and positive, $-x^3$ is large and negative, i.e. $f(x)$ is large and negative.

Hence the graph finishes in the 4th quadrant.

Sketch

To draw the sketch we have the following information:

The graph starts in the 2nd quadrant; has a minimum turning point at $(0,0)$; has a maximum turning point at $(\frac{2}{3}, \frac{4}{27})$; passes through the point $(1,0)$ and finishes in the 4th quadrant.

Hence sketch (figure 14),

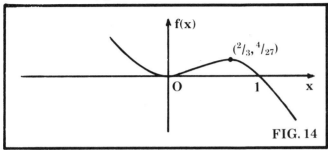

FIG. 14

Example 3

Sketch the graph of the function $f(x) = x^3(x-4)$.

(i) **Intersection with the axes**

$f(0) = 0$; hence point $(0,0)$ lies on the graph

$f(x) = 0 \Leftrightarrow x^3(x-4) = 0$

$\Leftrightarrow \quad x = 0$ (three times) or $x = 4$

$f(0) = 0$ and $f(4) = 0$. Hence the points $(0,0)$ and $(4,0)$ lie on the graph.

(ii) **Behaviour of $f(x)$ for large x**

$f(x) = x^4 - 4x^3$ and the behaviour depends on x^4.

If x is large and negative, x^4 is large and positive, i.e. $f(x)$ is large and positive.

Hence the graph starts in the 2nd quadrant.

If x is large and positive, x^4 is large and positive, i.e. $f(x)$ is large and positive.

Hence the graph finishes in the 1st quadrant.

(iii) **Stationary points**

$f(x) = x^3(x-4) = x^4 - 4x^3$

$f'(x) = 4x^3 - 12x^2 = 0 \Leftrightarrow 4x^2(x-3) = 0$

$\Leftrightarrow x = 0$ (twice) or $x = 3$

and $f(0) = 0$, $f(3) = 27(3-4) = -27$ are the stationary values and therefore the stationary points are $(0,0)$ and $(3, -27)$.

(iv) **Nature of the stationary points**

x	$\to$	0	$\to$	3	$\to$
$f'(x)$	$-$	0	$-$	0	$+$
		P.I.		Min.	
$f(x)$	↘	$\to$	↘	$\to$	↗

Hence $(0,0)$ is a point of inflexion on a *falling curve* and $(3, -27)$ is a minimum turning point.

(v) **Sketch**

To draw the sketch we have the following information:

The graph starts in the 2nd quadrant; has a point of inflexion at $(0,0)$; then a minimum turning point at $(3, -27)$; passes through the point $(4, 0)$ and finishes in the 1st quadrant.

Hence sketch (figure 15),

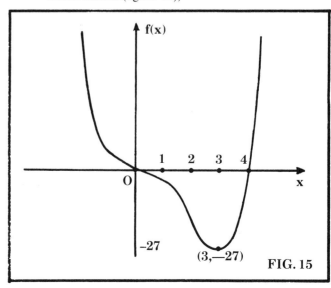

FIG. 15

3.3 SKETCHING THE GRAPHS OF FUNCTIONS WITHIN A CERTAIN INTERVAL

If we have to sketch the graph of a function f within a certain interval, say the closed interval $[a,b]$, we look for:

(i) (a) The point or points (if any) where the graph cuts the x-axis, such that $a \leq x \leq b$.

(b) The point or points (if any) where the graph cuts the y-axis if $a \leq x \leq b$.

(ii) The stationary values and stationary points provided that at the stationary points $a \leq x \leq b$.

(iii) The nature of the stationary values or points.

(iv) The values of the function at the end points $x = a$ and $x = b$, namely $f(a)$ and $f(b)$.

$f(a)$ and $f(b)$ may be the maximum or minimum values of the function f but not necessarily the maximum or minimum stationary values of the function.

Example 4

Sketch the graph of the function f defined by

$$f(x) = 2x^3 - 4x^2 + 12x$$

in the closed interval $[0,3]$, i.e. for $0 \leq x \leq 3$.

(i) **Intersection with the axes**

$f(0) = 0 \Rightarrow$ point $(0,0)$ lies on the graph

$f(x) = 0 \Leftrightarrow 2x^3 - 9x^2 + 12x = 0$

$\Leftrightarrow x(2x^2 - 9x + 12) = 0$

$\Rightarrow x = 0$ or $2x^2 - 9x + 12 = 0$

which has no roots

Thus $(0,0)$ is the only point at which the graph cuts the coordinate axes.

(ii) **Stationary values**

$$f(x) = 2x^3 - 9x^2 + 12x$$
$$f'(x) = 6x^2 - 18x + 12 = 0$$
$$\Leftrightarrow 6(x^2 - 3x + 2) = 0$$
$$\Leftrightarrow 6(x-1)(x-2) = 0$$
$$\Rightarrow x = 1 \quad \text{or} \quad x = 2$$

Thus $f(1) = 2 - 9 + 12 = 5$

$f(2) = 16 - 36 + 24 = 4$

are the stationary values.

The stationary points are $(1, 5)$, $(2, 4)$ and both $x = 1$ and $x = 2$ lie in the interval $[0, 3]$.

(iii) **Nature of the stationary values**

x	$\to$	1	$\to$	2	$\to$
$f'(x)$	+	0	−	0	−
$f(x)$	↗	Max.	↘	$\to$ Min.	↗

Thus $(1, 5)$ is a maximum stationary point and $(2, 4)$ is a minimum stationary point.

(iv) **Value of $f(x)$ at the end points $x = 0$ and $x = 3$**

$f(0) = 0$

$f(3) = 2.27 - 9.9 + 12.3 = 54 - 81 + 36 = 9$

(v) **Sketch**

To draw the sketch we have the following information:

The graph starts at $x = 0$ where $f(0) = 0$; has a maximum stationary point at $(1, 5)$; has a minimum stationary point at $(2, 4)$ and finishes at $x = 3$ where $f(3) = 9$.

Hence sketch (figure 16),

Note:

(i) The maximum value of f in the interval $[0, 3]$ is 9, but the maximum stationary value of f in $[0, 3]$ is 5.

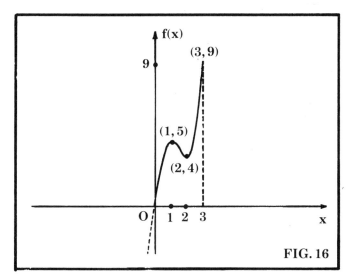

FIG. 16

(ii) The minimum value of f in the interval $[0,3]$ is 0, but the minimum stationary value of f in $[0,3]$ is 4.

ASSIGNMENT 3.1

Sketch the graphs of the following functions defined by $f(x)$.
[**Remember** to use some or all of the following points:
  (i) The points of intersection with the axes.
  (ii) The behaviour of $f(x)$ for large x.
  (iii) The stationary values and stationary points.
  (iv) The nature of the stationary values.]

1. $f(x) = 3x - x^3$
2. $f(x) = x^3 + x^2$
3. $f(x) = x^2 - x$
4. $f(x) = x^2 + 2x$
5. $f(x) = 3x^3$
6. $f(x) = x^2(x-6)$
7. $f(x) = x^4 - 2x^2 + 4$
8. $f(x) = 4 + 3x^2 - x^3$
9. $f(x) = x^3 - 3x + 2$
10. $f(x) = x^3(2-x)$

11. Show that $x+1$ and $x-2$ are factors of
$$f(x) = x^4 - 2x^3 - 3x^2 + 4x + 4$$
and find the other factors.
   Show that $x = 2$ gives a minimum turning point on the graph of the function defined by $f(x)$.
   Find the other two stationary points on the graph of the function and determine their nature.
   Sketch the graph of the function f.

Find the maximum and minimum value of the following functions defined by $f(x)$ in the closed intervals given and hence sketch the graph of each function.

12. $f(x) = x^2 + 2x$ $[-2, 0]$
13. $f(x) = 3 + 2x - x^2$ $[0, 3]$
14. $f(x) = x^2(3-x)$ $[0, 2]$
15. $f(x) = x^4 - x^2 - 2$ $[-\sqrt{2}, \sqrt{2}]$
16. $f(x) = 9x - x^3$ $[-3, 3]$
17. $f(x) = 4x^3 - x^4$ $[0, 3]$
18. $f(x) = \sqrt{x}(6-x)$ for $x > 0$ and where $\sqrt{x}$ means only the positive value is taken. Find the turning point and determine its nature.
   Sketch the graph of the function f in the closed interval $[0, 9]$.

PROBLEMS INVOLVING MAXIMA AND MINIMA 3.4

Example 5

An area of land has to be chosen for a small rectangular housing site. The planning department of the local authority have stipulated that the perimeter of the boundary should be no more than 2000 metres. What is the greatest rectangular area the builder has available for his site?

If lm is the length and bm the breadth of the site then

$$2l+2b = 2000 \Rightarrow l+b = 1000$$

Hence if he chooses, $l = 900$ and $b = 100$, the area equals 90,000 m², but if $l = 600$ and $b = 400$, the area equals 240,000 m².

So that by changing the length and breadth of the site but still keeping the boundary perimeter as 2000 m he can vary the area of the site. He has therefore to choose a length and breadth which will make the site area greatest (i.e. a maximum).

Let x metres be the length of the site then $(1000-x)$ metres is the breadth of the site, and the area in square metres is given by $A = x(1000-x)$ m². Thus he has to choose x so that A is a maximum.

Note in this problem $x > 0$ and

$$1000 - x > 0$$
$$\Rightarrow \quad 1000 > x$$
$$\Rightarrow \quad x < 1000$$

i.e. $\quad 0 < x < 1000$

Since the area A is a function defined by $A(x) = x(1000-x)$ then for A to be a maximum

$$\frac{dA}{dx} \text{ or } A'(x) = 0$$

$$A'(x) = 1000 - 2x = 0 \Leftrightarrow x = 500$$

To show $x = 500$ gives a maximum stationary value of A, use the table.

x	+	500	→
$A'(x)$	+	0	−
$A(x)$	↗	Max.	↘

Hence $x = 500$ gives a maximum area, i.e. the greatest area has a length of 500 m and a breadth of 500 m. Thus the greatest area is equal to 500×500 m², i.e. 250,000 m².

Example 6

An open water tank with rectangular sides and square base has to be built to hold 4000 litres of water. Find the dimensions of the tank so that the minimum amount of metal may be used in its construction (see figure 17).

$$4000 \text{ litres} = 4000 \times 10^3 \text{ cm}^3$$
$$= 4 \times 10^6 \text{ cm}^3$$
$$= 4 \text{ m}^3$$

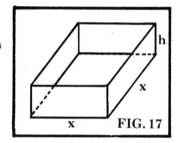

FIG. 17

Let x metres be the length of the tank and h metres the height of the tank.

Capacity of the tank is equal to $x^2 h$ m³. But capacity equals

$$4 \text{ m}^3 \Rightarrow 4 = x^2 h$$
$$\Rightarrow h = \frac{4}{x^2}$$

Area of metal used is

$$x^2 + 4xh = x^2 + 4x \cdot \frac{4}{x^2}$$

i.e. $$A(x) = x^2 + \frac{16}{x}$$

We have to choose x so that A is a minimum.

$$A'(x) = 2x - \frac{16}{x^2}$$

If the area is to be a minimum then $A'(x) = 0$

$$A'(x) = 0 \Leftrightarrow 2x - \frac{16}{x^2} = 0$$
$$\Rightarrow 2x^3 - 16 = 0$$
$$\Rightarrow x^3 = 8 \Rightarrow x = 2$$

and $\quad h = \dfrac{4}{x^2} = \dfrac{4}{4} = 1$

x	$\rightarrow$	2	$\rightarrow$
$A'(x)$	$-$	0	$+$
$A(x)$	$\searrow$	Min.	$\nearrow$

Hence $x = 2$ gives a minimum stationary value and the area of the metal used is a minimum when the length of the base is 2 metres and the height is 1 metre and then the tank will hold 4 cubic metres or 4000 litres of water.

ASSIGNMENT 3.2

1. A sheep's pen has to be constructed with 120 metres of wire fencing. Find the greatest area that can be enclosed by this amount of wire. [Let x metres be the length of wire fencing then $(60-x)$ metres is the breadth. Complete your reasoning as in worked example 5.]

2. The sum of two numbers is 56. Show that their product is a maximum when the numbers are equal.

3. A car park has to be fenced off with 500 metres of fencing, one wall acting as part of the car park. What is the greatest rectangular area of the car park?

4. An open water channel with a rectangular section has to be built to carry water, the metal available is of width 1 metre and has to be bent to form the channel. Find the length and breadth of the section so that the maximum amount of water may be carried by the channel.

5. Open rectangular bins with square bases and of volume 108 m^3 are to be made using a minimum of metal. If x m is the length of a side of the base and h m the height of each bin show that the total surface area of each bin is

$$x^2 + \frac{432}{x} \text{ m}^2$$

Calculate the dimensions of the bins so that the metal used is a minimum.

6. If the metal bins of question 5 are cylindrical, show that the area of metal used in constructing each bin is given by

$$A(r) = \frac{216}{r} + \pi r^2 \text{ m}^2$$

where r m is the radius of the base. Determine the radius of the base so that the surface area of the metal bins is a minimum.

(take $\pi = 3.14$)

7. The cost £P per day of manufacturing a certain article is given by the formula

$$P(n) = \frac{n^2}{3} + \frac{1152}{n}, n \neq 0$$

where n is the number of articles produced each day. Find the number of articles that must be produced each day to keep the cost at a minimum.

8. A and B are points on the coordinate axes, A has coordinates $(0, 9)$ and B$(12, 0)$. C(a, b) is any point on AB (see figure 18, over page). Show by proportion or otherwise that $4b + 3a = 36$.

CM and CN are parallel to the coordinate axes. Determine the coordinates of the point C, so that the area of rectangle OMCN is a maximum.

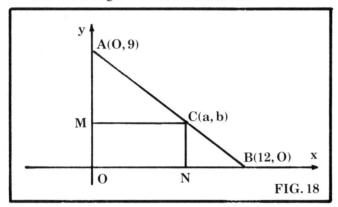

FIG. 18

9. ABCD is a rectangle of length 13 cm and breadth 7 cm. Points P, Q, R, S are taken on the sides of the rectangle such that AP = BQ = CR = DS = x cm (see figure 19).

Show that the area A of the parallelogram PQRS is given by

$$A(x) = 91 - 20x + 2x^2 \text{ cm}^2$$

Find the value of x for which this area is a minimum.

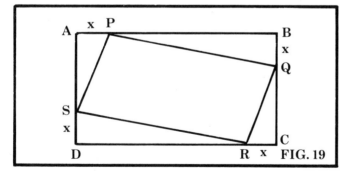

FIG. 19

10. Figure 20 shows a right circular cone inscribed in a sphere of radius R cm, and centre O.

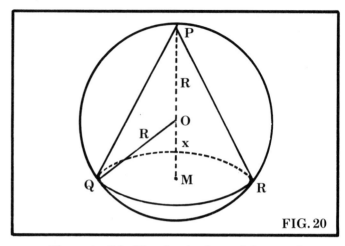

FIG. 20

The centre M of the circular base of the cone is x cm distant from O. Show that if V is the volume of the cone then,

$$V(x) = \tfrac{1}{3}\pi(R^3 + R^2x - Rx^2 - x^3) \text{ cm}^3$$

(*Note:* Find first QM^2 in terms of R and x. The volume of a cone is $\tfrac{1}{3}\pi r^2 h$ where r is the radius of the base and h is the height of the cone.)

Find in terms of R the value of x for which the volume is a maximum and find this maximum volume.

11. An open rectangular box with a square base of side x cm and height h cm is to be made from 675 cm² of material. Show that the height h can be expressed in terms of x in the form

$$h = \frac{675 - x^2}{4x}$$

and that the volume V of the box in terms of x can be written $V(x) = \tfrac{1}{4}x(675 - x^2) \text{ cm}^3$.

Determine the dimensions of the box so that the volume is a maximum and find this maximum volume.

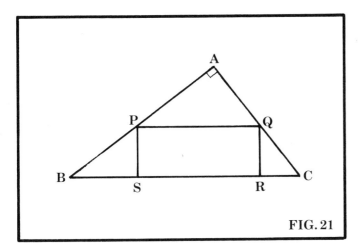

FIG. 21

12. ABC is a right-angled triangle with AB = 8 units, AC = 6 units and BC = 10 units.

   A rectangle PQRS is inscribed in the triangle, SR lying on BC, P on AB and Q on AC. PS = x units (see figure 21).

   Show by comparing each of the similar triangles BPS and QCR in turn with triangle ABC that BS = $\frac{4}{3}x$ and RC = $\frac{3}{4}x$.

   Show that the area of rectangle PQRS is given by $\frac{1}{12}x(120-25x)$ units^2.

   Find the value of x for which the area of this rectangle is a maximum and find this maximum area.

13. The sum of the height h and diameter d of a cylinder is 24 cm.

   Show that the volume V of the cylinder is

   $$\frac{\pi}{4}(24d^2 - d^3) \text{ cm}^3$$

   Find the value of d and h for which the volume of the cylinder will be a maximum and determine this maximum volume.

ASSIGNMENT 3.3

Objective type items testing Sections 3.1–3.4.

Instructions for answering these items are given on page 16.

Questions 1 to 4 refer to the graphs A to E (see figure 22).

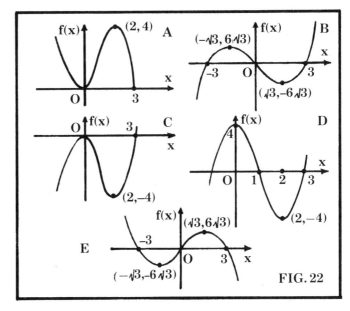

FIG. 22

1. Which graph is most likely to be that of the function f defined by $f(x) = x^2(x-3)$?

2. Which is the graph of the function f where
   $$f'(x) = 3(3-x^2)?$$

3. Which graph is most likely to represent a function f of the form
   $$f(x) = ax^3 - bx^2, \quad a, b \in N?$$

4. Which graph when reflected in the x-axis has image $F(x) = x(9-x^2)$?

155

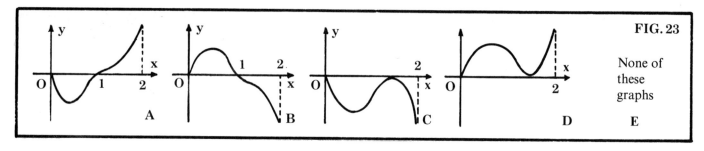

FIG. 23

5. The shape of the graph (see figure 23) of the function f defined by $f(x) = x(x-1)^3$ in the closed interval $[0,2]$ is

6. If the sum of the diameter d of the circular base and vertical height h of a cone is 27 cm then the volume of the cone may be written as

$$V(d) = \frac{\pi}{12}(27d^2 - d^3) \text{ cm}^3$$

The cone will have a maximum volume if
   A. $d = 0$
   B. $d = \frac{1}{4}h$
   C. $d = \frac{1}{2}h$
   D. $d = h$
   E. $d = 2h$

7. Figure 24 is a sketch of part of the graph of $y = x(x-2)$. The curve is symmetrical about
   A. $x = 0$
   B. $y = 0$
   C. $x = 1$
   D. $x = 2$
   E. $y = 2$

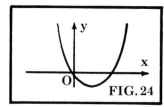

FIG. 24

8. A ball is thrown into the air and after t seconds it is h m above its point of projection where $h(t) = 4t(3-t)$. The time, in seconds, taken to reach its maximum height is
   A. $\frac{2}{3}$
   B. $\frac{3}{2}$
   C. 3
   D. 6
   E. 9

9. The curve with equation $y = x^3(x^2 - 1)$ has
   (1) A point of inflexion and two turning points.
   (2) Three points of intersection with the x-axis.
   (3) No point of inflexion.

10. The function $f: x \to x(x-2)^2$ is defined in the closed interval $[0,4]$.
    (1) The maximum value of the function occurs at an end point.
    (2) The minimum value occurs at $x = 2$ and $x = 0$.
    (3) The curve has no maximum stationary value.

11. (1) The graph of the function defined by $f(x) = x^n + 2$, n a positive integer, has a point of inflexion.
    (2) $n > 2$.

12. (1) 200 metres of fencing encloses a maximum rectangular area.
    (2) The area is a square of side 50 metres.

UNIT 4: RATES OF CHANGE

CONSTANT RATE OF CHANGE 4.1

A man $1\tfrac{3}{4}$ metres tall stands x metres from a lamp-post 7 metres high. How far is his shadow tip from the foot of the lamp-post (see figure 25)?

Let y metres be the required distance.
By proportion:

$$\frac{1\tfrac{3}{4}}{7} = \frac{y-x}{y}$$

$$\Rightarrow \quad \tfrac{1}{4} = \frac{y-x}{y}$$

$$\Rightarrow 4y - 4x = y$$
$$\Rightarrow \quad 3y = 4x$$
$$\Rightarrow \quad y = 1\tfrac{1}{3}x$$

Thus y defines a function f of x and $\dfrac{dy}{dx} = 1\tfrac{1}{3}$.

This means that if the distance x changes by any amount, the distance y changes in the same way but by $1\tfrac{1}{3}$ times the amount. That is any change (increase or decrease) in y is $1\tfrac{1}{3}$ metres for every metre increase or decrease in x.

For example, if x increases by 3 metres then y increases by 4 metres.

Here $\dfrac{dy}{dx}$ measures the rate of change of y with respect to x. This rate of change is constant and hence the change in y does not depend on x so that at any moment, any change in x is accompanied by a similar change in y but by $1\tfrac{1}{3}$ times the amount of change in x.

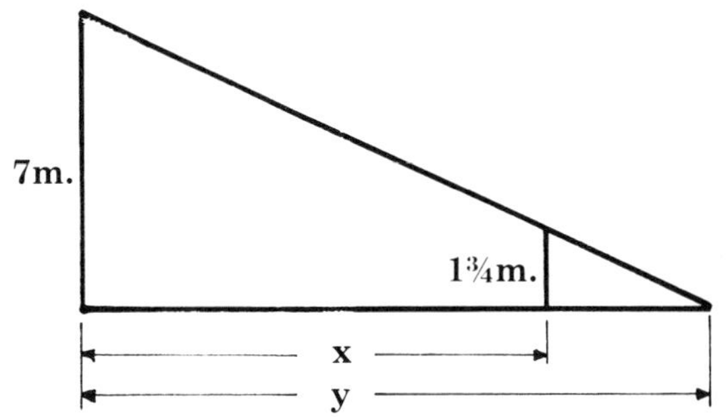

FIG. 25

Example 1

The length of the circumference of a circle is $C = 2\pi r$.

(i) Find the rate of change of the circumference with respect to the radius.

(ii) Find the increase in the circumference if the radius increases by 2 cm.

(i) $C = 2\pi r$, hence $\dfrac{dC}{dr} = 2\pi$.

Hence the rate of increase of the circumference with respect to r is 2π cm per cm.

(ii) Since the rate of change of the circumference with respect to the radius is constant, the increase in the circumference C is 2π cm for every centimetre increase in r.

Hence if the radius increases by 2 cm, the circumference increases by 4π cm.

4.3
4.2 WHEN THE RATE OF CHANGE IS NOT CONSTANT

Consider the area of a circle given by $A = \pi r^2$.

The rate at which the area A is changing with respect to the radius is $\dfrac{dA}{dr}$.

Hence the rate of change of A with respect to r is

$$\frac{dA}{dr} = 2\pi r$$

Here the rate of change of the area at any moment depends on the value of the radius r at that particular moment.

At the moment when $r = 2$ mm, the rate of change of area

$$\frac{dA}{dr} = 4\pi$$

Hence at the moment when $r = 2$ mm and only at this moment, the area is increasing by 4π mm² per mm.

Example 2

The surface area of a spherical balloon is $A = 4\pi r^2$, where r is the radius. Find the rate at which the surface area of the balloon is increasing with respect to the radius when the radius is 5 cm.

$$A = 4\pi r^2 \quad \text{and} \quad \frac{dA}{dr} = 8\pi r$$

when $r = 5$, $\dfrac{dA}{dr} = 8\pi \times 5 = 40\pi$

Hence at the moment when $r = 5$ cm, the area is increasing at the rate of 40π cm² per cm.

VARIATION WITH TIME

Examples of changes with time would be:

(i) a projectile fired vertically. Its height above the ground depends on the time it has been moving;

(ii) water pouring into a swimming pool at a certain rate. At what rate is the depth of the water in the pool increasing?

If $y = f(x)$, then y depends for its value on the value of x and $x = g(t)$, then x depends for its value on the value of t, where t measures time.

Hence $y = f[g(t)]$ and depends for its value on the value of t.

But $\dfrac{dy}{dx}$ measures the rate of change of y with respect to x and $\dfrac{dx}{dt}$ measures the rate of change of x with respect to t

and by the chain rule for differentiation
$$\frac{dy}{dt} = \frac{dy}{dx} \cdot \frac{dx}{dt}$$

Thus the rate of change of y with respect to t is $\frac{dy}{dx}$ multiplied by the rate of change of x with respect to t.

Note: If y is measured in cm and t in seconds, then $\frac{dy}{dt}$ is measured in cm per second.

Example 3

Drops of ink falling on a piece of absorbent paper cause a circular ink blot whose radius is increasing at the rate of 0·05 cm per second. At what rate is the area of the blot increasing at the moment when the radius is 1 cm?

If $r =$ radius and $A =$ area of the blot then $A = \pi r^2$.

We are given $\frac{dr}{dt} = 0·05$ cm per second and we want to find $\frac{dA}{dt}$

$A = \pi r^2$ and $\frac{dA}{dr} = 2\pi r$

$\qquad = 2\pi$ at the moment when $r = 1$ cm.

By the chain rule
$$\frac{dA}{dt} = \frac{dA}{dr} \cdot \frac{dr}{dt}$$
$$= 2\pi \times 0·05$$
$$= 0·314$$

Hence at the moment when $r = 1$, the area is increasing at the rate of 0·314 cm² per second.

Example 4

The height h metres of a projectile fired vertically after t seconds is given by the formula $h = 45t - 5t^2$. Calculate,

(i) its speed after 2 seconds.

(ii) its maximum height reached.

(iii) its total time of flight.

(i) Its speed is the rate of change of this height with respect to time, i.e. $\frac{dh}{dt}$.

Since $\frac{dh}{dt} = 45 - 10t = 45 - 20$ when $t = 2$.

Its speed after 2 seconds is 25 metres per second.

(ii) The height reached is a maximum

$\Leftrightarrow \quad \frac{dh}{dt} = 0$

$\Leftrightarrow 45 - 10t = 0$

$\Leftrightarrow \qquad 10t = 45$

$\Rightarrow \qquad t = 4\frac{1}{2}$

i.e. maximum height is reached after $4\frac{1}{2}$ seconds.

∴ maximum height reached is

$h = 45t - 5t^2$ when $t = 4\frac{1}{2}$

$= 45 \times \frac{9}{2} - 5 \times \frac{81}{4} = \frac{405}{2} - \frac{405}{4}$

$= \frac{405}{4} = 101\frac{1}{4}$ metres

Hence maximum height reached is $101\frac{1}{4}$ metres.

(iii) The maximum height is reached in $4\frac{1}{2}$ seconds. Hence total time of flight is 9 seconds.

ASSIGNMENT 4.1

1. The distance s metres travelled by a body in t seconds is given by $s = 1 \cdot 6t$. Calculate:
   (i) the distance gone by the body in 5 seconds.
   (ii) the rate of increase of the distance with respect to time.
   (iii) the distance gone in any 4 seconds.

2. The length of a rectangle is twice the breadth. Find the perimeter P of the rectangle in terms of the breadth b. Find the rate of change of the perimeter of the rectangle with respect to the breadth. If the breadth increases by 5 cm, by how much will the perimeter increase?

3. The length of a rectangle is twice the breadth. Find the area A of the rectangle in terms of the breadth b. Find the rate of change of the area of the rectangle with respect to the breadth b when the breadth is 3 cm.

4. The side of a square has length x cm. By how much is its area A increasing at the moment when $x = 4$?

   $\left(\text{You have to find } \dfrac{dA}{dx} \text{ when } x = 4, \text{ and your answer will be in cm}^2 \text{ per cm.}\right)$

5. The volume V of a spherical balloon is given by $V = \frac{4}{3}\pi r^3$ where r cm is the radius. Find the rate at which its volume is changing with respect to r when $r = 3$.

6. The length of a cuboid is twice the breadth and also twice the height. Write down the volume V in terms of the length l.
   Find the rate of change of the volume of the cuboid with respect to the length when the length is 2 cm.

7. The radius of a circular ink blot is increasing at the rate of 0·15 cm per second. At what rate is its area increasing at the moment when the radius is 5 cm?

8. Find the rate of change of the volume V of a spherical balloon with respect to its radius at the moment when the radius is equal to 6 cm.
   If the radius of the balloon is increasing at the rate of 4 cm per second, find the rate at which its volume is increasing when the radius is 6 cm.

9. The area of a circle is increasing at the rate of 3 cm^2 per second. Find,
   (i) the rate at which the radius is increasing at the moment when the radius is 6 cm.

   $\left(\text{Here } \dfrac{dA}{dt} = 3 \text{ cm}^2 \text{ per second and we require to find } \dfrac{dr}{dt}.\right)$

   (ii) the rate at which the circumference is increasing when the radius is 6 cm.

10. Air is leaking from a spherical balloon at the rate of 6 cm^3 per second. At what rate is its radius decreasing at the moment when the radius is 3 cm?

    $\left(\textit{Note:} \text{ Here } \dfrac{dV}{dt} = -6 \text{ cm}^3 \text{ per second.}\right)$

11. The volume V of a given quantity of gas is given by the formula $V = \dfrac{600}{p}$ where p is the pressure of the gas.
    If p increases at the rate of 0·25 kg per minute, by how much is V changing when $p = 15$?
    Is the volume increasing or decreasing when $p = 15$?

12. A vessel in the form of an inverted hollow cone is of height 5 metres and radius of base $\frac{1}{2}$ metre (see

figure 26). Water is poured into it. Show that when the depth of the water is x metres, the volume of water in the cone is given by,

$$V = \frac{\pi}{300} x^3 \text{ cubic metres.}$$

If water is poured in at the rate of 2 litres per second, find the rate at which the depth of water is increasing at the moment when the level of the water is halfway up the vessel.

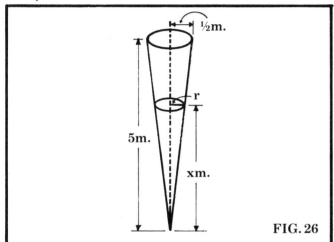

FIG. 26

13. A body dropped from a height of 800 metres moves according to the formula $s = 5t^2$, where s metres is the distance fallen after t seconds. Find,
    (i) the distance fallen in 5 seconds.
    (ii) the time taken to fall to the ground.
    (iii) the speed of the body when it has fallen for 5 seconds.

14. A ball thrown vertically upwards with a speed of 45 metres per second moves according to the equation $h = 45t - 5t^2$, where h is the height in metres above the starting point after t seconds. Find,
    (i) the speed after 4 seconds.
    (ii) the greatest height reached by the ball.

15. A body moving in a straight line is s metres from its starting point after t seconds, such that $s = 2t^2 + 5t - 1$. Find its speed,
    (i) after 4 seconds. (ii) after 10 seconds.

16. The volume V of a sphere of radius r is $V = \frac{4}{3}\pi r^3$ and the surface area S of the sphere is $S = 4\pi r^2$. Show that,
    (1) $\dfrac{dV}{dr} = S$.
    (2) $\dfrac{dV}{dt} = S\dfrac{dr}{dt}$.
    (3) $\dfrac{dV}{dt} = \dfrac{dS}{dt}$ when $r = 2$.

UNIT 5: THE INTEGRAL CALCULUS

INTEGRATION AS ANTI-DIFFERENTIATION — 5.1

Integration may be regarded as the inverse process of differentiation. This does not mean that in order to integrate a function we apply the rules of differentiation in reverse. However this application of the rules in reverse may be done in a limited number of cases.

$$f(x) = x^2 \Rightarrow f'(x) = 2x$$

hence if $\quad f'(x) = 2x \quad$ then $\quad f(x) = x^2$

$$f(x) = x^2 + 3 \Rightarrow f'(x) = 2x$$

hence if $\quad f'(x) = 2x \quad$ then $\quad f(x) = x^2 + 3$

$$f(x) = x^2 - 7 \Rightarrow f'(x) = 2x$$

hence if $\quad f'(x) = 2x \quad$ then $\quad f(x) = x^2 - 7$

We see from these three examples that

$$f(x) = x^2 + k, \quad k \in R \Rightarrow f'(x) = 2x$$

Thus given $f'(x) = 2x$, then $f(x) = x^2 + k$, where $k \in R$. k is called the **constant of integration** and is usually denoted by C.

Again given

$$\frac{dy}{dx} = 2x + 1 \quad \text{or} \quad f'(x) = 2x + 1$$

then $\quad y = x^2 + x + C \quad$ or $\quad f(x) = x^2 + x + C$

We can check that our answers are correct by differentiating y or $f(x) = x^2 + x + C$ to obtain

$$\frac{dy}{dx} \quad \text{or} \quad f'(x) = 2x + 1$$

Note: Any function which has $f'(x)$ as a derivative is called an integral of $f'(x)$ with respect to x.

Every integral of some function defined by $f(x)$ is contained in the formula $F(x) + C$.

This is written

$$\int f(x).dx = F(x) + C$$

$F(x) + C$ is called the general integral of $f(x)$; $F(x)$ is called the indefinite integral of $f(x)$; and C is called the constant of integration.

Thus $\int (2x+1).dx = x^2 + x + C$

which is the same as saying integrate the function $(2x+1)$ or saying, given $f'(x) = 2x+1$, what is $f(x)$?

Hence

$$\int f'(x).dx = \int (2x+1).dx = x^2 + x + C$$

$$\Rightarrow f(x) = x^2 + x + C$$

Rules for integration in some simple cases

Rule 1

If $\quad \dfrac{dy}{dx} \quad$ or $\quad f'(x) = x^n$

then $\qquad y = f(x) = \dfrac{x^{n+1}}{n+1} + C \quad (n \neq -1)$

or $\qquad \int x^n.dx = \dfrac{x^{n+1}}{n+1} + C \quad (n \neq -1)$

Example 1

Integrate (i) x^6 and (ii) $x^{-\frac{4}{3}}$.

(i) $\int x^6.dx = \dfrac{x^{6+1}}{6+1} + C = \dfrac{x^7}{7} + C.$

(ii) $\int x^{-\frac{4}{3}}.dx = \dfrac{x^{-\frac{4}{3}+1}}{-\frac{4}{3}+1} + C = \dfrac{x^{-\frac{1}{3}}}{-\frac{1}{3}} + C = -3x^{-\frac{1}{3}} + C.$

Rule 2

If $\quad \dfrac{dy}{dx} \quad$ or $\quad f'(x) = (ax+b)^n$

then $\qquad y = f(x) = \dfrac{(ax+b)^{n+1}}{a(n+1)} + C, n \neq -1$

or $\qquad \int (ax+b)^n.dx = \dfrac{(ax+b)^{n+1}}{a(n+1)} + C, n \neq -1$

Example 2

Integrate $(3-2x)^{-\frac{3}{2}}$.

$$\int (3-2x)^{-\frac{3}{2}}.dx = \dfrac{(3-2x)^{-\frac{1}{2}}}{-2(-\frac{1}{2})} + C = \dfrac{1}{\sqrt{(3-2x)}} + C$$

Rule 3

The integral of the sums of powers of x is equal to the sum of the integrals of the terms.

Example 3

If $\qquad f'(x) = x^3 + 6x^2 - 4x + 1$

then $\qquad f(x) = \dfrac{x^4}{4} + \dfrac{6x^3}{3} - \dfrac{4x^2}{2} + x + C$

$\qquad = \dfrac{x^4}{4} + 2x^3 - 2x^2 + x + C$

As a check, differentiate $f(x)$ to obtain $f'(x)$.

Example 4
Integrate $x^5 - x^3 + 2$.

Let $\quad f'(x) = x^5 - x^3 + 2$

then $\quad f(x) = \dfrac{x^6}{6} - \dfrac{x^4}{4} + 2x + C$

Or we can write it as

$$\int (x^5 - x^3 + 2).dx = \dfrac{x^6}{6} - \dfrac{x^4}{4} + 2x + C$$

Or as $\quad \dfrac{dy}{dx} = x^5 - x^3 + 2 \Rightarrow y = \dfrac{x^6}{6} - \dfrac{x^4}{4} + 2x + C$

Example 5
Given $f'(x) = x^{-\frac{3}{2}} - x^{-\frac{1}{2}}$ and that $f(4) = -1$, find $f(x)$.

$$f'(x) = x^{-\frac{3}{2}} - x^{-\frac{1}{2}} \Rightarrow f(x) = \dfrac{x^{-\frac{1}{2}}}{-\frac{1}{2}} - \dfrac{x^{\frac{1}{2}}}{\frac{1}{2}} + C$$

$$= -2x^{-\frac{1}{2}} - 2x^{\frac{1}{2}} + C$$

$$= -\dfrac{2}{\sqrt{x}} - 2\sqrt{x} + C$$

Thus $\quad f(4) = -\dfrac{2}{\sqrt{4}} - 2\sqrt{4} + C$

$\qquad\qquad = -1 - 4 + C$

But $\quad f(4) = -1 \Rightarrow -1 - 4 + C = -1$

$\qquad\qquad \Rightarrow \qquad C = 4$

Hence $\quad f(x) = -2x^{-\frac{1}{2}} - 2x^{\frac{1}{2}} + 4$

$\qquad\qquad = -\dfrac{2}{\sqrt{x}} - 2\sqrt{x} + 4$

ASSIGNMENT 5.1

1. Find $f(x)$ for the given derived functions. (Don't forget the constant of integration.)

   (i) $f'(x) = x^3$ \qquad (ii) $f'(x) = 4x^3 + x^2$

   (iii) $f'(x) = x^2 + x + 1$ \qquad (iv) $f'(x) = 3x^2 + 6x - 2$

   (v) $f'(x) = 2 - x$ \qquad (vi) $f'(x) = 15x^4 + 6x^2$

2. Given

   (i) $\dfrac{dy}{dx} = x + \dfrac{1}{\sqrt{x}}$ find y

   (ii) $\dfrac{dy}{dx} = x^{\frac{1}{2}} + x^{\frac{3}{2}}$ find y

   (iii) $\dfrac{dy}{dx} = x^3 + \dfrac{1}{x^2}$ find y

   (iv) $\dfrac{dy}{dx} = 3 - \tfrac{1}{2}x$ find y

Find,

3. $\int (x-2)^2 .dx$ \qquad 4. $\int (x^2 - 9).dx$

5. $\int \left(x - \dfrac{1}{x}\right)^2 .dx$ \qquad 6. $\int (x-1)(x+2).dx$

7. $\int \dfrac{x^3 + 1}{x^2}.dx$ \qquad 8. $\int x(x+1)(x-2).dx$

9. $\int \sqrt{x}.dx$ \qquad 10. $\int \sqrt{x}\left(1 - \dfrac{1}{x}\right).dx$

11. $\int (3x^{\frac{1}{2}} - x^{-\frac{1}{2}}).dx$ \qquad 12. $\int \left(x^2 - 1 - \dfrac{3}{x^2}\right).dx$

13. $\int x^{-\frac{1}{3}}(1-x)^2 .dx$ \qquad 14. $\int \left(\dfrac{1 + \sqrt{x}}{x^3}\right).dx$

15. $\int \tfrac{1}{2}(4-3x)^2 \, dx$

16. $\int \left(\dfrac{x^2-2x+1}{x^{\frac{1}{3}}}\right) dx$

17. $\int \left(x^2 - \dfrac{1}{x^4}\right)^2 dx$

18. $\int x^{-\frac{1}{4}}(1-2x)\, dx$

19. $\int \dfrac{1}{(3x-2)^3}\, dx$

20. $\int \dfrac{2}{\sqrt{(x-3)}}\, dx$

21. Integrate with respect to x.

   (i) $6x^2 - \dfrac{1}{x^2} + \dfrac{2}{x^3}$ (ii) $8x^3 + \dfrac{4}{x^3} - 3$

   (iii) $x^{\frac{3}{2}} - 4x^{-\frac{1}{2}} + 7$

22. If $\dfrac{dy}{dx} = 8x - \dfrac{4}{x^2}$ and $y = 14$ when $x = 2$, find y in terms of x.

23. If $f'(x) = 4x^3 - \dfrac{1}{x^3}$ and if $f(-2) = \tfrac{1}{4}$, find $f(x)$.

24. Given $\dfrac{dy}{dx} = x^{\frac{1}{2}} - x^{-\frac{1}{2}}$ and that $y = \tfrac{4}{3}$ when $x = 4$, find y in terms of x.

25. Given $f'(x) = x^{\frac{3}{2}} + x^{-\frac{1}{3}}$ and that $f(8) = 40$, find $f(x)$.

26. Integrate the following with respect to the variables,

   (i) $(2t+1)^7$ (ii) $(3x-4)^{-\frac{2}{3}}$

   (iii) $\dfrac{1}{\sqrt{(2x-1)}}$ (iv) $(3-4x)^5$

   (v) $\dfrac{1}{\sqrt{(1-4x)}}$ (vi) $(\tfrac{1}{2}t-4)^{-\frac{3}{2}}$

ASSIGNMENT 5.2

Objective type items testing Section 5.1.
Instructions for answering these items are given on page 16.

1. $\int x^{-\frac{3}{2}}\, dx$ equals

   A. $x^{\frac{1}{2}} + C$
   B. $2x^{\frac{1}{2}} + C$
   C. $\tfrac{1}{2}x^{\frac{1}{2}} + C$
   D. $-\tfrac{1}{2}x^{-\frac{1}{2}} + C$
   E. $-x^{-\frac{1}{2}} + C$

2. If $f'(x) = 5x^4 - 4x^3 + 3$, then $f(x)$ is equal to

   A. $x^5 - \tfrac{4}{3}x^4 + 3 + C$
   B. $x^5 - x^4 + 3 + C$
   C. $x^5 - x^4 + 3x + C$
   D. $\dfrac{x^5}{5} - x^4 + 3x + C$
   E. $\dfrac{x^5}{5} - \dfrac{x^4}{4} + 3x + C$

3. $\int (2\sqrt{x} + k)\, dx,\ k \in Q$, equals

   A. $\tfrac{3}{2}x^{\frac{3}{2}} + kx$
   B. $\tfrac{4}{3}x^{\frac{3}{2}} + C$
   C. $\tfrac{4}{3}x^{\frac{3}{2}} + kx + C$
   D. $\tfrac{4}{3}x^{\frac{3}{2}} + k + C$
   E. $x^{-\frac{1}{2}} + kx + C$

4. If $f'(x) = (1+6x)^{-\frac{1}{2}}$ then $f(x)$ equals

   A. $\tfrac{1}{3}(1+6x)^{\frac{1}{2}} + p$
   B. $-\tfrac{1}{3}(1+6x)^{\frac{1}{2}} + p$
   C. $x + 12x^{\frac{1}{2}} + p$
   D. $x + \dfrac{x^{\frac{1}{2}}}{3} + p$
   E. $2(1+6x)^{\frac{1}{2}} + p$

5. If $f'(x) = x^2 - 2x$ and $f(3) = 9$, then $f(x)$ equals

   A. $\dfrac{x^2}{3} - x^2 + 3$

   B. $\dfrac{x^3}{3} - x^2 - 3$

   C. $\dfrac{x^3}{3} - x^2$

   D. $\dfrac{x^3}{3} - x^2 - 9$

   E. $\dfrac{x^3}{3} - x^2 + 9$

6. The integral of $4x^3 - \dfrac{3}{x^4}$ is

   A. $4x^4 + \dfrac{1}{x^3} + C$

   B. $x^4 - \dfrac{1}{x^3} + C$

   C. $x^4 + \dfrac{3}{x^3} + C$

   D. $x^4 + \dfrac{1}{x^3} + C$

   E. $4x^4 + \dfrac{3}{x^3} + C$

7. $\displaystyle\int x(x-1)^3 \, dx$ equals

   A. $\dfrac{x^2(x-1)^3}{6} + C$

   B. $\left(\dfrac{x^3}{3} - \dfrac{x^2}{2}\right)^2 + C$

   C. $x^3 - 2x^2 + x + C$

   D. $-\dfrac{x^2}{2}(x-1)^3 + C$

   E. none of these

8. If $\dfrac{dy}{dx} = 2x - 1$ and $y = 4$ when $x = -1$, then y is equal to

   A. $x^2 - x$

   B. $x^2 - x + 2$

   C. $x^2 - x + 4$

   D. $x^2 - x + 6$

   E. none of these

9. If $f'(x) = 3x^2 - 4x$, then $f(x)$ can equal

   (1) $x^3 - 2x^2$

   (2) $4 - 2x^2 + x^3$

   (3) $x^3 - 2x^2 - 1$

10. If $\displaystyle\int f(x) \, dx = F(x) + C$ and $\displaystyle\int g(x) \, dx = G(x) + D$, then

    (1) $\displaystyle\int f(x) \, dx - \int g(x) \, dx = F(x) - G(x) + K$, where K is a constant.

    (2) $\displaystyle\int f(x) \cdot g(x) \, dx = F(x) \cdot G(x) + K$.

    (3) $\displaystyle\int \dfrac{f(x) \, dx}{g(x) \, dx} = \dfrac{F(x) + C}{G(x) + D}$.

11. (1) $f'(x) = x^3 + 2x$.

    (2) $f(x) = \dfrac{x^4}{4} + x^2 + 1$.

12. (1) $f(x) = x^3 - 2x^2$.

    (2) $f'(x) = 3x^2 - 4x$.

167

UNIT 6: DEFINITE INTEGRALS AND APPLICATIONS

THE DEFINITE INTEGRAL   6.1

The integral

$$\int_a^b y\,.\,dx = \int_a^b f(x)\,dx \quad \text{where} \quad y = f(x)$$

and which is read *the integral of y or f(x) from a to b* (i.e. from $x = a$ to $x = b$) is called a **definite integral**, a and b being the limits of integration, a the lower limit and b the upper limit.

If $$\int f(x)\,dx = F(x) + C$$

then $$\int_a^b f(x)\,dx = [F(x)]_a^b = F(b) - F(a)$$

Example 1

Evaluate the integral $\int_1^4 \left(6x + \dfrac{1}{x^2}\right) dx$

$$\int_1^4 \left(6x + \frac{1}{x^2}\right) dx = \int_1^4 (6x + x^{-2})\,dx = [3x^2 - x^{-1}]_1^4$$
$$= (48 - \tfrac{1}{4}) - (3 - 1)$$
$$= 47\tfrac{3}{4} - 2 = 45\tfrac{3}{4}$$

ASSIGNMENT 6.1

Evaluate the following definite integrals.

1. $\int_0^1 (3x-1)\,dx$
2. $\int_1^3 (x^2-2x)\,dx$
3. $\int_1^2 \frac{3}{x^2}\,dx$
4. $\int_0^1 (3x^2-4x+1)\,dx$
5. $\int_1^4 \left(6x - \frac{1}{x^2}\right)dx$
6. $\int_{\frac{1}{2}}^1 \frac{x^2+1}{x^2}\,dx$
7. $\int_0^2 (2x-1)^2\,dx$
8. $\int_{-1}^1 (8x^3-6x^2+2x)\,dx$
9. $\int_0^3 \sqrt{(1+t)}\,dt$
10. $\int_{-2}^2 (x+1)(2-x)\,dx$
11. $\int_1^3 \left(u^2 + \frac{2}{u^2}\right)du$
12. $\int_1^k \left(3x^2 - \frac{1}{x^2}\right)dx \quad k > 0$
13. $\int_{-4}^0 \frac{1}{\sqrt{(1-2u)}}\,du$
14. $\int_1^3 \left(t - \frac{1}{t}\right)^2 dt$
15. $\int_1^4 \left(\sqrt{x} + \frac{1}{\sqrt{x}}\right)dx$

16. Show that
$$\int_1^2 \left(2x + \frac{3}{x^2}\right)dx + \int_2^3 \left(2x + \frac{3}{x^2}\right)dx = \int_1^3 \left(2x + \frac{3}{x^2}\right)dx$$

17. If $\int_0^k x^{\frac{1}{3}}\,dx = 12$, find the value of k.

18. Determine p given that $\int_1^p x^{-\frac{1}{2}}\,dx = 6$.

6.2 THE DEFINITE INTEGRAL AS AN AREA

If $y = f(x)$ then the integral

$$\int_a^b f(x)\,dx = \int_a^b y\,dx$$

represents the area bounded by the curve $y = f(x)$, the x-axis and the lines $x = a$ and $x = b$ (see figure 27).

The shaded area $A = \int_a^b f(x)\,dx = \int_a^b y\,dx$.

Example 2

Find the area between the curve with equation
$$y = 3 + 2x - x^2,$$
the x-axis and the lines $x = 0$ and $x = 2$. Figure 28 shows the curve $y = 3 + 2x - x^2$ with the lines $x = 0$ and $x = 2$. The shaded part is the required area.

$$\text{Required area} = \int_0^2 y\,dx$$
$$= \int_0^2 (3 + 2x - x^2)\,dx$$

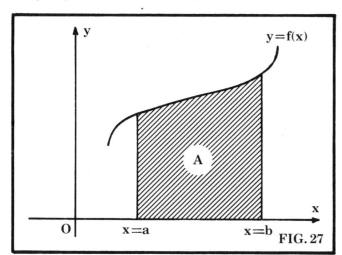

FIG. 27

$$= \left[3x+x^2-\frac{x^3}{3}\right]_0^2$$
$$= (6+4-\tfrac{8}{3})-0$$
$$= \tfrac{22}{3} \text{ units}^2$$

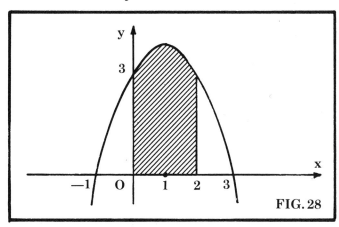

FIG. 28

Example 3

Find the area between the curve $y = 2x(x^2-1)$ and the x-axis,

(i) from $x = -1$ to $x = 0$

(ii) from $x = 0$ to $x = 1$.

Figure 29 shows part of the curve $y = 2x(x^2-1)$ cutting the x-axis at $x = -1$, $x = 0$ and $x = 1$. The shaded parts A_1 and A_2 are the required areas.

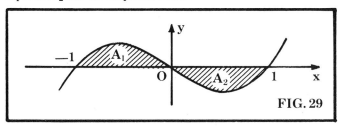

FIG. 29

(i) $A_1 = \int_{-1}^{0} 2x(x^2-1)\,dx = \int_{-1}^{0} (2x^3-2x)\,dx$

$= \left[\dfrac{x^4}{2} - x^2\right]_{-1}^{0} = 0 - [\tfrac{1}{2}-1] = \tfrac{1}{2}$

(ii) $A_2 = \int_{0}^{1} 2x(x^2-1)\,dx = \int_{0}^{1} (2x^3-2x)\,dx$

$= \left[\dfrac{x^4}{2} - x^2\right]_{0}^{1} = (\tfrac{1}{2}-1)-0 = -\tfrac{1}{2}$

This value of the area is negative because the area A_2 is below the x-axis, where y is negative.

Hence, *Note* that the area between the curve, the x-axis and the lines $x = -1$ and $x = 1$ is $A_1 + A_2$ where the **numerical** value of A_2 is taken.

i.e. $A_1 + A_2 = \tfrac{1}{2}+\tfrac{1}{2} = 1 \text{ unit}^2$

In general if a curve cuts the x-axis at more than two points then the value of the area between each pair of points must be calculated separately and the **numerical** values added together.

Hence if possible, always draw a rough sketch of your areas before performing the integration.

Example 4

Show in a diagram the area represented by

$$\int_0^2 \sqrt{(4-x^2)}\,dx$$

and hence evaluate this integral.

$$\text{Area} = \int_a^b y \cdot dx = \int_0^2 \sqrt{(4-x^2)}\,dx$$

Hence $y = \sqrt{(4-x^2)}$
$\Rightarrow y^2 = 4-x^2$
$\Rightarrow x^2+y^2 = 4$

171

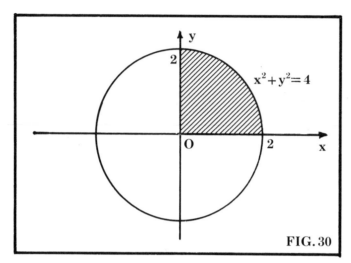

FIG. 30

and this is the equation of a circle, centre the origin and of radius 2 units.

Hence we can draw the sketch and the shaded region shows the area represented by

$$\int_0^2 \sqrt{(4-x^2)}\,dx$$

namely a quadrant of the area of the circle $x^2+y^2=4$, since the limits are from $x=0$ to $x=2$ (see Figure 30). But the area of a circle of radius 2 is equal to 4π and the area of a quadrant of the circle $= \pi$.

Hence $\int_0^2 \sqrt{(4-x^2)}\,dx = \pi$

ASSIGNMENT 6.2

In questions 1–4 make a rough sketch of the graph of each function defined by $f(x)$ and shade in the area bounded by the graph, the x-axis and the given lines. Calculate the area bounded by the graph, the x-axis and the given lines.

1. $f(x) = 2x^2$ from $x = 0$ to $x = 2$.
2. $f(x) = 2x^2$ from $x = -1$ to $x = 2$.
3. $f(x) = 4x^3$ from $x = 0$ to $x = 2$.
4. $f(x) = 4x^3$ from $x = -2$ to $x = 2$. (*Note:* part of the required area is *below* the x-axis.)
5. Sketch the curve $y^2 = x$ and find the area bounded by the curve, the x-axis and the lines $x = 1$ and $x = 4$.
6. Find the area under the curve $y = \sqrt{(1+x)}$ from $x = 0$ to $x = 3$.
7. Find the area between the x-axis, the curve $y = x(x-2)$ and the lines,
   (i) $x = -2$ and $x = 0$
   (ii) $x = 0$ and $x = 2$
8. Sketch roughly the curve $y = x(x-1)(x-2)$ and find the total area enclosed by it and the x-axis.
9. Show in a diagram the area represented by

$$\int_{-3}^{3} \sqrt{(9-x^2)}\,dx$$

and hence evaluate this integral.

10. Show in a diagram the area represented by

$$\int_0^2 \sqrt{(4x-x^2)}\,dx$$

and hence evaluate this integral.

11. Show that

$$\int_{-2}^{2} x(x^2-4)\,dx = 0$$

but that the total area between the curve $y = x(x^2-4)$, the x-axis and the lines $x = -2$ and $x = 2$ is 8 square units.

12. Sketch the curve $y = x(2-x)$ for values of x from 0 to 4. Find the area between the curve, the x-axis and the lines $x = 0$ and $x = 4$.

    Evaluate $\int_0^4 x(2-x)\,dx$

    Explain your result.

13. Find the area under the parabola $y = 6 + x - x^2$ between the lines $x = -1$ and $x = 2$.

14. Sketch the curve $y = (x+2)(1-x)$ and show that the area bounded by the curve and the x-axis between $x = -2$ and $x = 1$ is divided by the y-axis in the ratio $20:7$.

15. Find the point of intersection in the first quadrant of the line $x + y = 3$ and the curve $y^2 = 4x$. Find the area in the first quadrant enclosed by the x-axis, the curve $y^2 = 4x$ and the lines $x = 0$ and $x + y = 3$.

THE AREA BETWEEN TWO CURVES

Let the two curves have equations $y_1 = f(x)$ and $y_2 = g(x)$, and intersect at $x = a$ and $x = b$ (see figure 31).

A_1 = the area bounded by the x-axis, the curve $y_1 = f(x)$ and the lines $x = a$ and $x = b$.

A_2 = the area bounded by the x-axis, the curve $y_2 = g(x)$ and the lines $x = a$ and $x = b$.

Required area (i.e. area between the curves) = $A_1 - A_2$.

But $\quad A_1 = \int_a^b y_1\,dx \quad$ and $\quad A_2 = \int_a^b y_2\,dx$

Hence the

Required area = $\int_a^b (y_1 - y_2)\,dx = \int_a^b [f(x) - g(x)]\,dx$

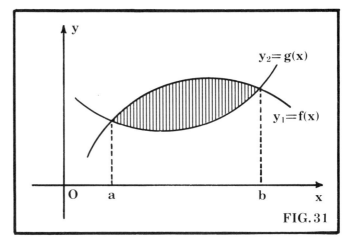

FIG. 31

Example 5

Find the area between the parabola $y = x^2 - 3x + 3$ and the line $y = x$.

The curve $y = x^2 - 3x + 3$ intersects the line $y = x$

$\Leftrightarrow x^2 - 3x + 3 = x$

$\Leftrightarrow x^2 - 4x + 3 = 0$

$\Leftrightarrow (x-1)(x-3) = 0$

$\Rightarrow \qquad x = 1 \text{ or } 3$, i.e. at the points $(1, 1)$ and $(3, 3)$

For $y = x^2 - 3x + 3$:

when $x = 0$, $y = 3$, $\Rightarrow$ the point $(0, 3)$ lies on the parabola;

when $y = 0$, $x^2 - 3x + 3 = 0$ and this has no roots in x.

Turning point:

$$\frac{dy}{dx} = 2x - 3 = 0 \Leftrightarrow x = 1\tfrac{1}{2}$$

When $x = 1\tfrac{1}{2}$, $y = \tfrac{9}{4} - \tfrac{9}{2} + 3 = \tfrac{3}{4}$, hence there is a minimum turning point at $(\tfrac{3}{2}, \tfrac{3}{4})$.

Hence sketch in figure 32 where the shaded part shows the required area.

6.3

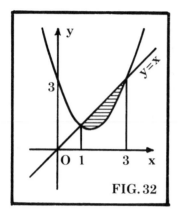

FIG. 32

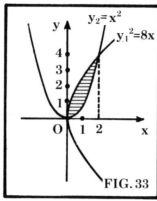

FIG. 33

Required area $= \int_1^3 (y_{\text{line}} - y_{\text{curve}})\,dx$

$= \int_1^3 [x - (x^2 - 3x + 3)]\,dx$

$= \int_1^3 (x - x^2 + 3x - 3)\,dx$

$= \int_1^3 (4x - x^2 - 3)\,dx$

$= \left[2x^2 - \dfrac{x^3}{3} - 3x \right]_1^3$

$= (18 - 9 - 9) - (2 - \tfrac{1}{3} - 3)$

$= 0 - (-1 - \tfrac{1}{3})$

$= 1\tfrac{1}{3}$ units^2

Example 6

Find the area between the curves $y = x^2$ and $y^2 = 8x$.

Note: $y = x^2$ is the parabola with vertex at $(0,0)$ and the y-axis as axis of symmetry.

$y^2 = 8x$ or $x = \tfrac{1}{8}y^2$ is also a parabola with vertex at $(0,0)$ but with the x-axis as axis of symmetry.

The curves $y = x^2$ and $y^2 = 8x$ intersect

$\Leftrightarrow \quad x^4 = 8x$

$\Leftrightarrow \quad x^4 - 8x = 0$

$\Leftrightarrow x(x^3 - 8) = 0$

$\Rightarrow \qquad x = 0 \quad \text{or} \quad x = 2$, i.e. they intersect at $(0,0)$ and $(2,4)$.

Hence sketch in figure 33 where the shaded part shows the required area.

Required area $= \displaystyle\int_0^2 (y_1 - y_2)\,dx$ where

$y_1 = \sqrt{(8x)}$ and $y_2 = x^2$

$= \int_0^2 [\sqrt{(8x)} - x^2]\,dx$

$= \int_0^2 (\sqrt{8} \cdot x^{\frac{1}{2}} - x^2)\,dx$

$= \left[\dfrac{\sqrt{8} \cdot x^{\frac{3}{2}}}{\frac{3}{2}} - \dfrac{x^3}{3} \right]_0^2$

$= \left[\dfrac{2\sqrt{8} \cdot x^{\frac{3}{2}}}{3} - \dfrac{x^3}{3} \right]_0^2$

$= \left(\dfrac{2\sqrt{8} \cdot 2\sqrt{2}}{3} - \dfrac{8}{3} \right) - 0$

$= \tfrac{16}{3} - \tfrac{8}{3} = \tfrac{8}{3}$ units^2

ASSIGNMENT 6.3

In questions 1–7 find the area between the given curves and the given straight lines.

(Remember to draw a rough sketch and to find the x-coordinates of the points of intersection of the curve and line.)

1. $y = x^2$ and $y = 2x$.

2. $y = x^2$ and $y = 4$.

3. $y = x(2-x)$ and $y = x$.

4. $y = x(x-4)$ and $y = -2x$.

5. $y = (4-x)(1+x)$ and $y = x+4$.

6. $y = x^3$ and $y = 4x$ (*Note:* that part of the area is *below* the x-axis).

7. $y = -x(5+x)$ and $y = -3x$.

8. Sketch the parabolas $y = 1+x^2$ and $y = 3-x^2$ and calculate the area between them.

9. Find the points of intersection of the parabolas $y = x(5-x)$ and $y = x^2+2$. Sketch the curves and find the area enclosed by them.

10. Find the area between the parabolas $y = x^2$ and $y^2 = 27x$. Show that the line $y = 3x$ bisects the area between the parabolas.

11. Show that the parabolas $2y^2 = x$ and $x^2 = 4y$ intersect at the points $(0,0)$ and $(2,1)$ and hence show that the area between them is $\tfrac{2}{3}$ square units.

12. Figure 34 shows part of the curve with equation $4y^2 = x(4-x)^2$. Find the area enclosed by the loop.

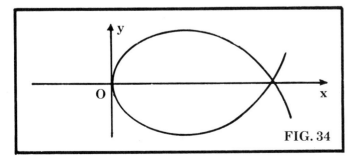

FIG. 34

13. Find the area bounded by the parabola $y = 8+x-x^2$ and the line $y+x = 5$. Show that the y-axis divides this area in the ratio $5:27$.

THE VOLUME OF A SOLID OF REVOLUTION 6.4

If the plane area bounded by the curve $y = f(x)$, the x-axis and the lines $x = a$ and $x = b$ is revolved round the x-axis, then it generates a solid which is symmetrical about the x-axis. The curve $y = f(x)$ between $x = a$ and $x = b$ sweeps out the surface of this solid (see figure 35).

The solid which is produced in this way is called a **solid of revolution.**

The volume of the solid is given by,

$$V = \int_a^b \pi y^2 \, dx \quad \text{where} \quad y = f(x)$$

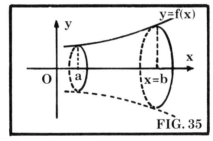
FIG. 35

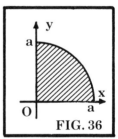
FIG. 36

Example 7

The area in the first quadrant bounded by the x-axis, the y-axis and the circle $x^2+y^2 = a^2$ is revolved round the x-axis. Find the volume of the solid formed (see figure 36).

The area in the first quadrant of the circle $x^2+y^2 = a^2$ is shown in the diagram. When this area is revolved round the x-axis the solid formed is a hemisphere whose volume V is given by

$$V = \int_0^a \pi y^2 \, dx \quad \text{where} \quad y^2 = a^2 - x^2$$

175

$$\therefore V = \int_0^a \pi(a^2-x^2)\,dx = \pi\int_0^a (a^2-x^2)\,dx$$
$$= \pi\left[a^2x - \frac{x^3}{3}\right]_0^a = \pi\left[a^3 - \frac{a^3}{3}\right]$$
$$= \pi \cdot 2\frac{a^3}{3} = \tfrac{2}{3}\pi a^3$$

This gives the volume of a hemisphere of radius a.

Example 8

Find the volume of the solid whose surface is generated by revolving round the x-axis the curve $y = x\sqrt{(1-x^2)}$ from $x = 0$ to $x = 1$. (No sketch is needed here.)

$$y = x\sqrt{(1-x^2)} \Rightarrow y^2 = x^2(1-x^2)$$

hence

$$\text{Volume of the solid} = \int_0^1 \pi y^2\,dx = \pi\int_0^1 x^2(1-x^2)\,dx$$
$$= \pi\int_0^1 (x^2 - x^4)\,dx$$
$$= \pi\left[\frac{x^3}{3} - \frac{x^5}{5}\right] = \pi(\tfrac{1}{3} - \tfrac{1}{5})$$
$$= \frac{2\pi}{15}\text{ units}^3$$

6.5 THE VOLUME OF THE SOLID GENERATED BY REVOLVING ROUND THE x-AXIS THE AREA BETWEEN TWO CURVES (OR A CURVE AND A LINE)

This is best illustrated by an example.

Example 9

The segment of the parabola $y = x(4-x)$ cut off by the line $y = x$ is revolved about the x-axis. Calculate the volume of the solid formed. (A sketch is useful here.)

The parabola and the line intersect

$$\Leftrightarrow \quad x = x(4-x)$$
$$\Leftrightarrow x - x(4-x) = 0$$
$$\Leftrightarrow \quad x(x-3) = 0$$
$$\Rightarrow \quad x = 0 \text{ or } 3$$

Thus the curve cuts the x-axis at $(0,0)$ and $(4,0)$.

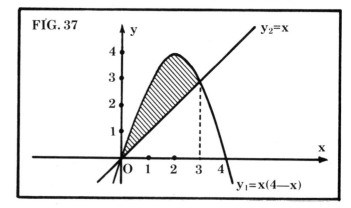

FIG. 37

Hence sketch in figure 37, and the shaded area is revolved round the x-axis. We want to find the volume of the solid generated by the shaded area.

If V_1 = the volume of the solid generated by the area under the parabola between the limits $x = 0$ and $x = 3$ then

$$V_1 = \int_0^3 \pi y_1^2\,dx$$

If V_2 = the volume of the solid generated by the area under the line and between the limits $x = 0$ and $x = 3$, then

$$V_2 = \int_0^3 \pi y_2^2 \cdot dx$$

Hence the required volume $= \int_0^3 \pi(y_1^2 - y_2^2)\,dx$

$$= \int_0^3 \pi[(4x-x^2)^2 - x^2]\,dx$$

$$= \pi \int_0^3 (16x^2 - 8x^3 + x^4 - x^2)\,dx$$

$$= \pi \int_0^3 (x^4 - 8x^3 + 15x^2)\,dx$$

$$= \pi \left[\frac{x^5}{5} - 2x^4 + 5x^3\right]_0^3$$

$$= \pi\left(\frac{243}{5} - 162 + 135\right)$$

$$= \pi\left(\frac{243}{5} - 27\right)$$

$$= \frac{108\pi}{5}\ \text{units}^3$$

Note: In general if the area between the curves $y_1 = f(x)$ and $y_2 = g(x)$ is rotated about the x-axis, the volume of the solid formed is

$$\pi \int_a^b \{[f(x)]^2 - [g(x)]^2\}\,dx$$

where a and b are the x-coordinates of the points of intersection of y_1 and y_2.

ASSIGNMENT 6.4

In questions 1–6 calculate the volumes of the solids formed when the areas bounded by the x-axis, the given curves and the lines are revolved through 360° about the x-axis.

1. $y = 3x^2$ and the lines $x = 0$ and $x = 1$.
2. $y = \tfrac{1}{3}x^2$ and the lines $x = 0$ and $x = 3$.
3. $y^2 = 2x$ and the lines $x = 2$ and $x = 4$.
4. $y = x - \dfrac{1}{x}$ and the lines $x = 1$ and $x = 2$.
5. $xy = 4$ and the lines $x = \tfrac{1}{2}$ and $x = 4$.
6. $y = \dfrac{1}{x-1}$ and the lines $x = -1$ and $x = -3$.
7. The area between the parabola $y = x(2-x)$ and the line $y = x$ is revolved once round the x-axis. Calculate the volume of the solid formed.
8. Figure 38 shows part of the curve with equation $y^2 = x(2-x)^2$. If the loop of this curve is rotated once round the x-axis, find the volume of the solid formed.

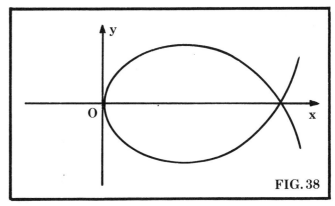

FIG. 38

9. The area between the parabola $y = (x+1)(4-x)$ and the line $y = 4$ is revolved round the x-axis. Find the volume of the solid formed.
10. Show that the circle $x^2 + y^2 = 4$ and the circle $x^2 + y^2 - 4x = 0$ with centre $(2, 0)$ and of radius 2 intersect at the point $(1, \sqrt{3})$ in the first quadrant. If the area between the two circles is revolved round the x-axis, show by drawing a sketch of the circles that the volume of the solid formed is given by

$$V = \pi \int_0^1 (4x - x^2)\,dx + \pi \int_1^2 (4 - x^2)\,dx$$

and hence evaluate this volume.

6.6 AREAS, VOLUMES AND THE y-AXIS

(a) The area bounded by the curve $x = g(y)$, the y-axis and the lines $y = a$ and $y = b$ is given by,

$$A = \int_a^b x \, dy \quad \text{where} \quad x = g(y) \text{ (see figure 39)}$$

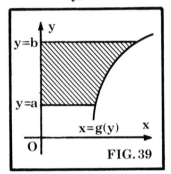

FIG. 39

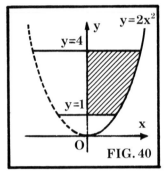

FIG. 40

y-axis and the lines $y = a$ and $y = b$ is revolved round the y-axis then it generates a solid which is symmetrical about the y-axis (see figure 41).

The volume of the solid is given by,

$$V = \int_a^b \pi x^2 \, dy \quad \text{where} \quad x = g(y)$$

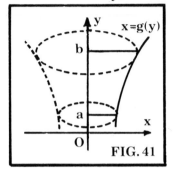

FIG. 41

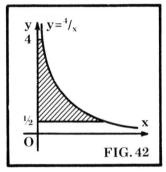

FIG. 42

Example 10

Find the area in the first quadrant enclosed by the curve $y = 2x^2$, the y-axis and the lines $y = 1$ and $y = 4$ (see figure 40).

$$\text{Required area} = \int_1^4 x \, dy$$

where
$$y = 2x^2 \Rightarrow x = \frac{\sqrt{y}}{\sqrt{2}}$$

defines the part of the curve in the first quadrant.

Hence area $= \int_1^4 \frac{\sqrt{y}}{\sqrt{2}} dy = \frac{1}{\sqrt{2}} \int_1^4 y^{\frac{1}{2}} \, dy = \frac{1}{\sqrt{2}} [\tfrac{2}{3} y^{\frac{3}{2}}]_1^4$

$= \frac{2}{3\sqrt{2}} [4^{\frac{3}{2}} - 1] = \frac{2}{3\sqrt{2}} (8 - 1)$

$= \frac{14}{3\sqrt{2}} = \frac{7\sqrt{2}}{3} \text{ units}^2$

(b) If the plane area bounded by the curve $x = g(y)$, the

Example 11

The area between the curve $y = \dfrac{4}{x}$, the y-axis and the lines $y = \tfrac{1}{2}$ and $y = 4$ is rotated round the y-axis. Calculate the volume of the solid formed (see figure 42).

$$\text{Volume} = \int_{\frac{1}{2}}^4 \pi x^2 \, dy \quad \text{where} \quad x^2 = \frac{16}{y^2}$$

$= \pi \int_{\frac{1}{2}}^4 \frac{16}{y^2} dy = 16\pi \int_{\frac{1}{2}}^4 y^{-2} \, dy$

$= 16\pi [-y^{-1}]_{\frac{1}{2}}^4 = -16\pi \left[\frac{1}{y}\right]_{\frac{1}{2}}^4$

$= -16\pi [\tfrac{1}{4} - 2] = -16\pi(-\tfrac{7}{4})$

$= 28\pi \text{ units}^3$

ASSIGNMENT 6.5

In questions 1–4, make a sketch of the curve and shade in the area in the first quadrant bounded by the y-axis, the curve and the given lines. Calculate this area and the volume obtained when this area is revolved once round the y-axis.

1. $y = 2x^2$ and the lines $y = 0$ and $y = 2$.
2. $y^2 = 4x$ and the lines $y = 0$ and $y = 4$.
3. $x^2 = y(2-y)^2$ and the lines $y = 0$ and $y = 2$.
4. $y^2 = \dfrac{1}{x}$ and the lines $y = \frac{1}{2}$ and $y = 1$.
5. Find the points of intersection of the curve $x = y(4-y)$ and the y-axis. Find the area enclosed by the curve and the y-axis and the volume obtained when this area is revolved about the y-axis.
6. Figure 43 shows part of the parabola $y^2 = 8 - 4x$. The shaded area lies between the parabola and the line $x = 1$.
   Find the coordinates of A and B.
   Calculate the volume generated when this area revolves round the y-axis.
7. Figure 44 shows part of the curve $y^2 = x^3$ between $x = 0$ and $x = 4$.
   Find the volume of the solid formed,
   (i) when the area between the curve and the x-axis from $x = 0$ to $x = 4$ is revolved round the x-axis.
   (ii) when the area in the first quadrant between the curve and the y-axis from $x = 0$ and $x = 4$ is revolved once round the y-axis.
8. Find the points of intersection of the parabola $y^2 = 4x$ and the line $y = x$. Find the volume of the solid obtained when the area between the parabola and line is revolved,
   (i) round the x-axis.
   (ii) round the y-axis.
9. Sketch the parabolas $y = 1 + x^2$ and $y = 3 - x^2$ showing the points of intersection with each other and the points of intersection with the y-axis.
   If the area between the parabolas is revolved once round the x-axis, show that the volume of the solid formed is given by
   $$V_1 = \pi \int_{-1}^{1} (3 - x^2)^2 \, dx - \pi \int_{-1}^{1} (1 + x^2)^2 \, dx$$
   and evaluate this volume.
   If the area between the parabolas is revolved round the y-axis, show that the volume of the solid formed is given by
   $$V_2 = \pi \int_{1}^{2} (y - 1) \, dy + \pi \int_{2}^{3} (3 - y) \, dy$$
   and evaluate this volume.
10. Figure 45 shows the ellipse with equation
    $$\dfrac{x^2}{9} + \dfrac{y^2}{4} = 1$$
    cutting the x-axis at $(3, 0)$ and $(-3, 0)$, and the y-axis at $(0, 2)$ and $(0, -2)$.

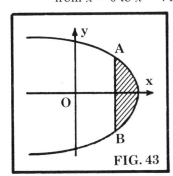

FIG. 43

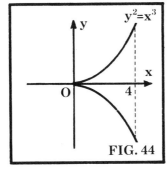

FIG. 44

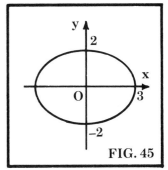

FIG. 45

Find the volume of the ellipsoid formed by revolving the ellipse about the *x*-axis.

Is the same volume obtained by revolving the ellipse round the *y*-axis? Check by finding the volume of the solid formed when the ellipse is revolved round the *y*-axis.

ASSIGNMENT 6.6

Objective type items testing Sections 6.1–6.5.
Instructions for answering these items are given on page 16.

1. The value of $\int_1^4 \frac{1}{\sqrt{x}} \, dx$ equals

   A. $\frac{1}{2}$
   B. $1\frac{1}{2}$
   C. 2
   D. $\frac{14}{3}$
   E. 6

2. The value of $\int_{-2}^{2} 3x^2 \, dx$ is

   A. 0
   B. 8
   C. 12
   D. 16
   E. 24

3. The area of the shaded region in figure 46 is

   A. $\int_0^8 x^3 \, dx$

   B. $\int_0^8 \sqrt[3]{(x)} \, dx$

   C. $\int_0^2 \sqrt[3]{(x)} \, dx$

   D. $\int_0^2 x^3 \, dx$

   E. $\int_2^8 x^3 \, dx$

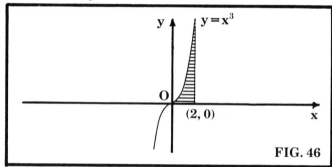

FIG. 46

4. The value of $\int_0^a (a-2x)^2 \, dx$ is

   A. $\frac{2a^3}{3}$
   B. $-a^3$
   C. $-\frac{2a^3}{3}$
   D. $\frac{a^3}{3}$
   E. $-\frac{a^3}{3}$

5. The shaded area in the diagram (figure 47) is revolved once round the x-axis. The volume of the solid formed is

A. $\int_1^2 \pi x^3 \, dx$

B. $\int_1^2 \pi x^{3/2} \, dx$

C. $\int_1^2 \pi(x^3 - x^2) \, dx$

D. $\int_1^2 \pi x^2 \, dx$

E. $\int_1^2 \pi x \, dx$

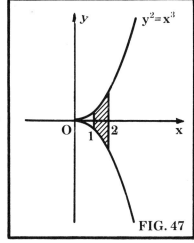

FIG. 47

6. The area of the shaded region in figure 48 is

A. $\int_1^2 y^2 \, dy$

B. $\int_1^4 x^2 \, dx$

C. $\int_1^4 y^{1/2} \, dy$

D. $\int_1^2 x^2 \, dx$

E. $\int_1^2 x^{1/2} \, dx$

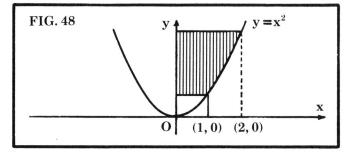

7. The volume $\int_1^2 \pi \cdot \dfrac{dx}{x^2}$ is obtained by revolving round the x-axis between $x = 1$ and $x = 2$ the curve with equation,

A. $yx^2 = 1$

B. $y^2 = x$

C. $y^2 x = 1$

D. $xy = 1$

E. $x^2 = y$

8. The volume of the solid obtained when the shaded area in figure 49 is rotated through one complete revolution about the y-axis is given by

A. $\pi \int_0^q y^{1/4} \, dy$

B. $\pi \int_0^q y \, dy$

C. $\pi \int_0^q y^{1/2} \, dy$

D. $\pi \int_0^p y^{1/2} \, dy$

E. $\pi \int_0^p x^8 \, dx$

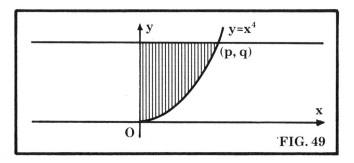

FIG. 49

9. The value of the integral
$$\int_{-a}^{0} f(x)\,dx + \int_{0}^{a} f(x)\,dx = 0 \qquad a > 0$$
is true for which of the following

(1) $\int_{-1}^{0} x^2\,dx + \int_{0}^{1} x^2\,dx$

(2) $\int_{-1}^{0} x^3\,dx + \int_{0}^{1} x^3\,dx$

(3) $\int_{-1}^{0} \frac{1}{x^2}\,dx + \int_{0}^{1} \frac{1}{x^2}\,dx$

10. Which of the following integrals is/are equal in value to the integral
$$\int_{1}^{2} \left(x - \frac{1}{x^2}\right) dx$$

(1) $\int_{-1}^{-2} \left(x - \frac{1}{x^2}\right) dx$

(2) $\int_{-1}^{-2} \left(x + \frac{1}{x^2}\right) dx$

(3) $\int_{1}^{2} \left(t - \frac{1}{t^2}\right) dt$

11. (1) $\int_{a}^{b} f(x)\,dx = -\int_{b}^{a} f(x)\,dx$

(2) $f(x) = 1$

12. (1) $\int_{0}^{3} \sqrt{(9-x^2)}\,dx = \frac{9\pi}{4}$

(2) The integral represents the area of a quadrant of a circle of radius 3 units and centre the origin.

UNIT 7: TRIGONOMETRICAL FUNCTIONS

DERIVATIVE OF THE SINE AND COSINE FUNCTIONS

(a) If $f(x) = \sin x$ then $f'(x) = \cos x$.

(b) If $f(x) = \cos x$ then $f'(x) = -\sin x$.

To prove these we use the following facts.

(i) $f'(x) = \lim\limits_{h \to 0} \dfrac{f(x+h) - f(x)}{h}$

(ii) $\sin C - \sin D = 2 \cos \dfrac{C+D}{2} \sin \dfrac{C-D}{2}$

or (iii) $\cos C - \cos D = -2 \sin \dfrac{C+D}{2} \sin \dfrac{C-D}{2}$ $(C > D)$

(iv) $\lim\limits_{h \to 0} f(x) \cdot g(x) = \lim\limits_{h \to 0} f(x) \cdot \lim\limits_{h \to 0} g(x)$

(v) $\lim\limits_{h \to 0} \dfrac{\sin h}{h} = 1$.

(a) If $f(x) = \sin x$

$$f'(x) = \lim_{h \to 0} \frac{\sin(x+h) - \sin x}{h} \quad \text{using (i)}$$

$$= \lim_{h \to 0} \frac{2 \cos(x + \tfrac{1}{2}h) \sin \tfrac{1}{2}h}{h} \quad \text{using (ii)}$$

$$= \lim_{h \to 0} \cos(x + \tfrac{1}{2}h) \cdot \frac{\sin \tfrac{1}{2}h}{\tfrac{1}{2}h}$$

$$= \lim_{h \to 0} \cos(x + \tfrac{1}{2}h) \cdot \lim_{h \to 0} \frac{\sin \tfrac{1}{2}h}{\tfrac{1}{2}h} \quad \text{using (iv)}$$

$$= \cos x \cdot 1 \quad \text{using (v)}$$

$$= \cos x$$

Hence if $y = \sin x$, $\dfrac{dy}{dx} = \cos x$.

(b) If $f(x) = \cos x$

$$f'(x) = \lim_{h \to 0} \frac{\cos(x+h) - \cos x}{h} \quad \text{using (i)}$$

$$= \lim_{h \to 0} \frac{-2 \sin(x + \tfrac{1}{2}h) \cdot \sin \tfrac{1}{2}h}{h} \quad \text{using (iii)}$$

$$= \lim_{h \to 0} -\sin(x + \tfrac{1}{2}h) \cdot \frac{\sin \tfrac{1}{2}h}{\tfrac{1}{2}h}$$

$$= \lim_{h \to 0} -\sin(x + \tfrac{1}{2}h) \cdot \lim_{h \to 0} \frac{\sin \tfrac{1}{2}h}{\tfrac{1}{2}h} \quad \text{using (iv)}$$

$$= -\sin x \cdot 1$$

$$= -\sin x$$

Hence if $y = \cos x$, $\dfrac{dy}{dx} = -\sin x$.

7.1

7.2 FURTHER USE OF THE CHAIN RULE

These two derivatives along with the chain rule enables us to find the derivatives of further sine and cosine functions.

Example 1

Find the derivative of (i) $\sin 4x$
(ii) $\cos(3x+2)$
(iii) $\sin^3 x$

(i) Let $y = \sin 4x$
$\quad = \sin u$, where $u = 4x$

Then $\quad \dfrac{dy}{du} = \cos u \quad$ and $\quad \dfrac{du}{dx} = 4$

Hence $\quad \dfrac{dy}{dx} = \dfrac{dy}{du} \cdot \dfrac{du}{dx} = 4 \cos u = 4 \cos 4x$

Hence if $f(x) = \sin 4x$ then $f'(x) = 4 \cos 4x$.

(ii) Let $y = \cos(3x+2)$
$\quad = \cos u$ where $u = 3x+2$

Then $\quad \dfrac{dy}{du} = -\sin u \quad$ and $\quad \dfrac{du}{dx} = 3$

Hence $\quad \dfrac{dy}{dx} = \dfrac{dy}{du} \cdot \dfrac{du}{dx} = -3 \sin u = -3 \sin(3x+2)$

Hence if $f(x) = \cos(3x+2)$ then $f'(x) = -3 \sin(3x+2)$.

(iii) Let $y = \sin^3 x$
$\quad = u^3$ where $u = \sin x$

Then $\quad \dfrac{dy}{du} = 3u^2 \quad$ and $\quad \dfrac{du}{dx} = \cos x$

Hence $\quad \dfrac{dy}{dx} = \dfrac{dy}{du} \cdot \dfrac{du}{dx} = 3u^2 \cos x = 3 \sin^2 x \cdot \cos x$

Hence if $f(x) = \sin^3 x$ then $f'(x) = 3 \sin^2 x \cdot \cos x$.

Hence note that if $a \neq 0$ and $a, b \in Q$ then

(i) $\dfrac{d}{dx} \sin ax = a \cos ax \quad$ and $\quad \dfrac{d}{dx} \cos ax = -a \sin ax$

(ii) $\dfrac{d}{dx} \sin(ax+b) = a \cos(ax+b)$

and $\dfrac{d}{dx} \cos(ax+b) = -a \sin(ax+b)$

(iii) $\dfrac{d}{dx}(\sin^n x) = n \sin^{n-1} x \cdot \cos x$

and $\dfrac{d}{dx}(\cos^n x) = -n \cos^{n-1} x \cdot \sin x$

Example 2

For what values of x in the domain $0 \leq x \leq 2\pi$ does the function $f(x) = 1 - \sin x$ increase?

$$f(x) = 1 - \sin x \Rightarrow f'(x) = -\cos x$$

If $f(x)$ is increasing then $f'(x) > 0$

$\Rightarrow -\cos x > 0$
$\Rightarrow \cos x < 0$

But $\quad \cos x < 0 \quad$ if $\quad \dfrac{\pi}{2} < x < \dfrac{3\pi}{2}$

hence $f(x)$ increases for $\dfrac{\pi}{2} < x < \dfrac{3\pi}{2}$.

ASSIGNMENT 7.1

Find the derivatives of each of the following.

1. $6 \cos x$
2. $-3 \sin x$

3. $\sin x - 2\cos x$
4. $4\cos x - 3\sin x$
5. $x^2 + \sin x$
6. $1 - \tfrac{1}{2}\cos x$
7. $\sin 3x$
8. $\cos 4x$
9. $\sin(3x+2)$
10. $\cos(4x-5)$
11. $\cos(2-3x)$
12. $\sin(1-x)$
13. $\sin^4 x$
14. $\cos^5 x$
15. $(2-\sin x)^2$
16. $\sin 2x + \cos 4x$
17. $\cos x + \cos^2 x$
18. $\sin x + \cos^2 x$
19. $(1+\cos x)^2$
20. $\sin 3x - \cos^4 x$

21. Show that $\dfrac{d}{dx}(\sin^2 x - \cos^2 x) = 2\sin 2x$.

22. Find the equation of the tangent to the curve,
   (i) $y = \sin 2x$ at the point where $x = \dfrac{\pi}{2}$
   (ii) $y = \cos^2 x$ at the point where $x = \dfrac{\pi}{4}$
   (iii) $y = \sin x + \cos x$ at the point where $x = \pi$.

23. Show that the function f defined by $f(x) = 1 - \cos x$ never decreases for $0 < x < \pi$.

24. For what values of x in $0 < x < 180$ is the function defined by $1 + \sin x$ decreasing.

25. For what values of x in $0 \leq x \leq 180$ is the function f defined by $f(x) = \cos x + \sin x$ increasing.

26. Determine for what values of x for $0 \leq x \leq 360$ the function f defined by $f(x) = \cos^2 x$ has stationary values. Determine the nature of the stationary values given by these values of x.

27. For what values of x in $0 < x < 360$ has the function f defined by $f(x) = \sin 2x$ stationary values. Determine the nature of these stationary values.

28. Show that the function f defined by $f(x) = \sin x - \cos x$ has two turning points for $0 \leq x \leq 2\pi$. Find the corresponding turning values and determine their nature.

29. Discuss the stationary values of the function f defined by $f(x) = \cos x + \tfrac{1}{2}\sin 2x$ in the interval $0 \leq x \leq 2\pi$.

INTEGRALS OF TRIGONOMETRICAL FUNCTIONS 7.3

Since

(i) $\dfrac{d}{dx}\sin x = \cos x$ then $\displaystyle\int \cos x \, dx = \sin x + C$

(ii) $\dfrac{d}{dx}\cos x = -\sin x$ then $\displaystyle\int \sin x \, dx = -\cos x + C$

(iii) $\dfrac{d}{dx}\sin ax = a\cos ax$ then $\displaystyle\int \cos ax \, dx = \dfrac{1}{a}\sin ax + C$

(iv) $\dfrac{d}{dx}\cos ax = -a\sin ax$

then $\displaystyle\int \sin ax \, dx = -\dfrac{1}{a}\cos ax + C$

(v) $\dfrac{d}{dx}\sin(ax+b) = a\cos(ax+b)$

then $\displaystyle\int \cos(ax+b) \, dx = \dfrac{1}{a}\sin(ax+b) + C$

(vi) $\dfrac{d}{dx}\cos(ax+b) = -a\sin(ax+b)$

then $\displaystyle\int \sin(ax+b) \, dx = -\dfrac{1}{a}\cos(ax+b) + C$

Example 3

Integrate the functions (i) $\sin 4x$

(ii) $\cos(3x-2)$

(iii) $\sin(1-4x)$

(i) $\int \sin 4x \, dx = -\frac{1}{4}\cos 4x + C$ (using iv above)

(ii) $\int \cos(3x-2) \, dx = \frac{1}{3}\sin(3x-2) + C$ (using v above)

(iii) $\int \sin(1-4x) \, dx = -\frac{1}{-4}\cos(1-4x) + C$

(using vi above)

$= \frac{1}{4}\cos(1-4x) + C$

Example 4

Find the area enclosed by the curve $y = \cos x$ and the x-axis between $x = 0$ and $x = \frac{\pi}{2}$. Find also the volume of the solid formed when this area is revolved once round x-axis (see figure 50).

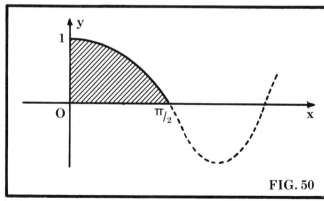

FIG. 50

The required area is shown in the diagram.

$\text{Area} = \int_0^{\pi/2} y \, dx = \int_0^{\pi/2} \cos x \, dx$

$= [\sin x]_0^{\pi/2}$

$= 1 \text{ unit}^2$

$\text{Volume} = \int_0^{\pi/2} \pi y^2 \, dx = \pi \int_0^{\pi/2} \cos^2 x \, dx$

But $\cos 2x = 2\cos^2 x - 1 \Rightarrow 2\cos^2 x = 1 + \cos 2x$

$\Rightarrow \cos^2 x = \frac{1}{2}(1 + \cos 2x)$

Hence $\text{volume} = \pi \int_0^{\pi/2} \frac{1}{2}(1 + \cos 2x) \, dx$

$= \frac{\pi}{2}\left[x + \frac{\sin 2x}{2}\right]_0^{\pi/2}$

$= \frac{\pi}{2}\left[\left(\frac{\pi}{2} + \frac{\sin \pi}{2}\right) - 0\right]$

$= \frac{\pi^2}{4} \text{ unit}^3$ (since $\sin \pi = 0$)

ASSIGNMENT 7.2

Integrate the following functions,

1. $\cos 2x$
2. $\sin 5x$
3. $\sin(3x+2)$
4. $\cos(4x+1)$
5. $\cos(-x+2)$
6. $\sin(2-3x)$
7. $-\cos 6x$
8. $-\sin 4x$
9. $-\cos(x+1)$
10. $-2\sin(4x+2)$
11. Express $\cos^2 x$ in terms of $\cos 2x$ and hence find $\int \cos^2 x \, dx$.

12. Express $\sin^2 x$ in terms of $\cos 2x$ and hence find
$$\int \sin^2 x \, dx.$$

Evaluate the following integrals,

13. $\displaystyle\int_0^{\pi/2} \cos x \, dx$

14. $\displaystyle\int_0^{\pi} \sin x \, dx$

15. $\displaystyle\int_{-\pi/2}^{\pi/2} \sin 2x \, dx$

16. $\displaystyle\int_0^{\pi/6} \cos 3x \, dx$

17. Express $\sin x \cdot \cos 2x$ as the difference of two sines and hence evaluate $\displaystyle\int_0^{\pi/2} \sin x \cdot \cos 2x \, dx$.

18. Show on a diagram the area represented by the integrals
    (i) $\displaystyle\int_0^{\pi/2} \sin 2x \, dx$
    (ii) $\displaystyle\int_0^{\pi/2} \cos 2x \, dx$
    Calculate the values of these areas.

19. Express $(\sin\theta + \cos\theta)^2$ in the form $a + \sin p\theta$, where $a, p \in R$ and hence evaluate $\displaystyle\int_0^{\pi/2} (\sin\theta + \cos\theta)^2 \, d\theta$.

20. For what values of x in $0 \leq x \leq \dfrac{\pi}{2}$ is $\cos 2x = 0$? Find the total area between the curve $y = \cos 2x$, the x-axis and the lines $x = 0$ and $x = \dfrac{\pi}{2}$.

21. Find the area between the curve $y = x + \cos 2x$, the x-axis and the lines $x = 0$ and $x = \dfrac{\pi}{2}$.

22. Sketch on the same diagram the graphs of the functions f and g defined by $f(x) = \sin x$ and $g(x) = \cos x$ between $x = 0$ and $x = \dfrac{\pi}{2}$.

Show that the area bounded by the curves and the x-axis between $x = 0$ and $x = \dfrac{\pi}{2}$ is given by,
$$A = \int_0^{\pi/4} \sin x \, dx + \int_{\pi/4}^{\pi/2} \cos x \, dx$$
and hence evaluate this area.

If this area is revolved once round the x-axis, show that the volume of the solid formed is given by,
$$V = \pi \int_0^{\pi/4} \sin^2 x \, dx + \pi \int_{\pi/4}^{\pi/2} \cos^2 x \, dx,$$
and hence by expressing $\sin^2 x$ and $\cos^2 x$ in terms of $\cos 2x$, evaluate this volume.

23. Figure 51 shows part of the curves with equations $y = \sin x$ and $y = \sin 2x$.
    Show that the point P of intersection of the two curves has x-coordinate equal to $\dfrac{\pi}{3}$.
    Calculate the area of the shaded part.

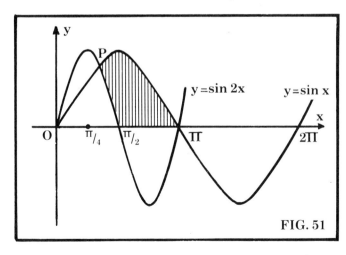

FIG. 51

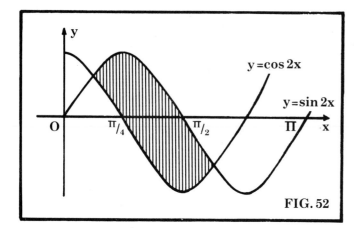

FIG. 52

ASSIGNMENT 7.3

Objective type items testing sections 7.1–7.3.
Instructions for answering these items are given on page 16.

1. If $f(x) = \sin(2x+5)$, then $f'(x)$ is equal to
   A. $\cos(2x+5)$
   B. $\cos 2(2x+5)$
   C. $2\cos(2x+5)$
   D. $-2\cos(2x+5)$
   E. $-\tfrac{1}{2}\cos(2x+5)$

2. $\int \cos(3x-1)\,dx$ is equal to
   A. $-\tfrac{1}{3}\sin(3x-1)+C$
   B. $\tfrac{1}{3}\sin(3x-1)+C$
   C. $3\sin(3x-1)+C$
   D. $-\sin(3x-1)+C$
   E. $\sin(3x-1)+C$

3. The derivative of $\sin^2 2x$ is
   A. $2\sin 2x \cos 2x$
   B. $2\cos^2 2x$
   C. $2\sin^2 2x$
   D. $4\sin 2x \cos 2x$
   E. $\sin 2x \cos 2x$

4. The area of the shaded region in figure 52 is
   A. $2\int_{\pi/4}^{\pi/2}(\cos 2x - \sin 2x)\,dx$
   B. $2\int_{\pi/4}^{\pi/2}(\sin 2x - \cos 2x)\,dx$
   C. $2\int_{\pi/8}^{\pi/2}\sin 2x\,dx - 2\int_{\pi/8}^{\pi/4}\cos 2x\,dx$
   D. $2\int_{\pi/8}^{\pi/2}(\cos 2x - \sin 2x)\,dx$
   E. $2\int_{0}^{\pi/2}\sin 2x\,dx - 2\int_{0}^{\pi/4}\cos 2x\,dx$

5. The area of the shaded region in figure 53 is
   A. $\int_0^1 (1-\sin x)\,dx$
   B. $\int_0^1 (x-\sin x)\,dx$
   C. $\int_0^{\pi/2}(x-\sin x)\,dx$
   D. $\int_0^{\pi}(\sin x - x)\,dx$
   E. $\int_0^{\pi/2}(1-\sin x)\,dx$

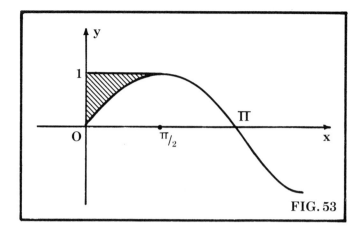

FIG. 53

6. The number of stationary values the function f defined by $f(x) = \cos\left(2x + \dfrac{\pi}{2}\right)$ in the domain $0 \leqq x \leqq 2\pi$ has is

   A. 0
   B. 1
   C. 2
   D. 3
   E. 4

7. The function f defined by $f(x) = \sin x + x$ has a stationary value when x is equal to

   A. $-\dfrac{\pi}{2}$
   B. 0
   C. $\dfrac{\pi}{2}$
   D. π
   E. $\dfrac{3\pi}{2}$

8. The function f defined by $f(\theta) = \cos \theta$ decreases in the domain $0 \leqq \theta \leqq 2\pi$, for

   A. $0 < \theta < \dfrac{\pi}{2}$ only
   B. $0 < \theta < \pi$
   C. $\dfrac{\pi}{2} < \theta < \pi$
   D. $\pi < \theta < \dfrac{3\pi}{2}$
   E. all values of θ in the domain $0 < \theta < 2\pi$.

9. The graph of the function f defined by
$$f(x) = \sin x + \cos x$$
has turning values at x equal to

   (1) $\dfrac{\pi}{4}$
   (2) $\dfrac{3\pi}{4}$
   (3) $\dfrac{5\pi}{4}$

10. If $f(\theta) = \cos^2 \theta$ then $f'(\theta)$ is equal to

    (1) $\sin 2\theta$
    (2) $-2 \sin \theta \cos \theta$
    (3) $-\sin 2\theta$

11. (1) $\displaystyle\int_0^a \sin x \, dx = \int_0^a \cos x \, dx$
    (2) $a = \dfrac{\pi}{2}$

12. (1) $f(x) = \sin 2x - \cos 2x$
    (2) $f'(\pi) = 2$

ASSIGNMENT 7.4

SUPPLEMENTARY EXAMPLES

1. Differentiate from first principles

   (i) $2x^2 - 3$ (ii) $\dfrac{1}{x^3}$ (iii) $\dfrac{2}{x^2}$

   Differentiate the following with respect to x

2. $x^{\frac{3}{2}} + 3x^{\frac{1}{2}} + \dfrac{2}{x^2},\ x > 0$ 3. $\dfrac{1}{\sqrt{(2-2x^3)}},\ x < 1$

4. $\dfrac{x^2 + 2x}{\sqrt{x}},\ x > 0$ 5. $\dfrac{x^2 + 2x - 7}{\sqrt{x}},\ x > 0$

6. (i) $\dfrac{1}{x^4},\ x \neq 0$ (ii) $x^3 - 7x + \dfrac{1}{3x},\ x \neq 0$

7. p is a function of v defined by $pv^{-2.5} = c$ where c is a constant. If $\dfrac{dp}{dv} = -18$ when $v = 9$ find the value of c.

8. Differentiate with respect to t

   (i) $(3t^2 - 1)(2 - t)$ (ii) $\dfrac{t+1}{t^3},\ t \neq 0$

   (iii) $\left(2t - \dfrac{1}{t}\right)^2,\ t \neq 0$ (iv) $t^k - \dfrac{1}{t^k},\ t > 0$

9. Find the value of t for which $\dfrac{dy}{dt} = 0$, when

   (i) $y = 3t^2 - t - 2$ (ii) $y = t^4 - 2t^3$

10. If $f: x \to 2x + \dfrac{1}{2x},\ x \neq 0$, find the value of x such that $f': x \to 0$.

11. Differentiate

    (i) $\dfrac{1 - x^3}{\sqrt{x}}$ (ii) $x - \sin 2x$ (iii) $\cos^2 x + \sin^2 x$

12. If $p = \cos 3t - 2 \sin t$ find $\dfrac{dp}{dt}$ when $t = \dfrac{\pi}{4}$.

13. Show that the tangent at the point where $x = -2$ on the curve given by $y = x(x-1)(x+3)$ has gradient 1, and find the equation of the tangent at the point where $x = -2$. Find also the coordinates of the other point on the curve where the tangent has a gradient of 1.

14. Find the equation of the tangent to the curve $y = 2x^3 - 8x$ at the point where $x = 1$. Find the co-ordinates of the point on the curve at which a parallel tangent can be drawn.

15. Find the gradient of the tangent to the curve given by $y = (1 - 3x)(2 + x)$ at the points where it cuts the x-axis. Find also the equation of the tangent to the curve at the point where it cuts the y-axis.

16. Find the coordinates of the points on the curve $y = x + \dfrac{1}{x}$ at which the tangents to the curve have gradient -3.

17. Find the coordinates of the points on the curve $y = \dfrac{x^2}{2} - \dfrac{1}{x}$ at which the tangents to the curve are parallel to the x-axis.

18. For what range of values of x is the function f defined by $f(x) = x^2 - 6x - 7$ negative?
    For what interval is the function f both negative and decreasing?

19. Prove that the function f defined by $f(x) = x - \dfrac{1}{x},\ x \neq 0$, increases for all values of x, $(x \neq 0)$.

20. For what values of x is the function f defined by $f(x) = 3x^3 - 6x^2 + 3x - 4$ decreasing as x increases?

21. For what values of t in the interval $0 < t < \pi$ is the function f defined by $f(t) = \frac{1}{2}\cos 2t - \cos t$ increasing?

22. Find the turning points on the curve with equation $y = x^2(4 - 4x - 2x^2)$, stating their nature. Obtain the points of intersection of the curve with the axes and hence sketch the curve.

23. Find the coordinates of the turning points on the curve with equation $y = x^3 - 6x^2 + 9x$ and investigate the nature of each. Draw a sketch of the curve.

24. If the sum of the length and diameter of a cylindrical drum is 6 metres, find in cubic metres the maximum volume of the drum.

25. Find the maximum and minimum value of the function f defined by $f: t \to \frac{1}{2}\cos 2t + \sin t$ in the interval $0 < t < \pi$.

26. A wire 4 metres long is cut into two pieces. One piece is made into a circle and the other into a square. Show that the diameter of the circle and the side of the square are equal in length when the total area enclosed by the two pieces is a minimum.

27. Figure 54 shows a rectangle PQRS inscribed in a semicircle centre O and radius r. If angle POQ = θ show that the area A of rectangle PQRS may be written $A = r^2 \sin 2\theta$. Hence find the dimensions of this rectangle if its area is to be a maximum. Write down this maximum area.

28. A window for a large church is in the shape of a rectangle surmounted by an equilateral triangle. If the perimeter of the window is to be 33 metres find the dimensions of the window that will admit the maximum amount of light. (Leave answers in surd form.)

29. The circumference of a circle is increasing at the rate of 2 centimetres per second. Find the rate at which the

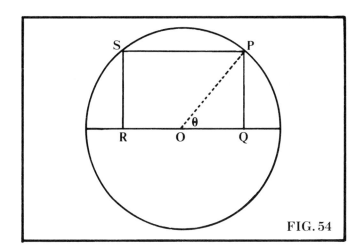

FIG. 54

area is increasing at the instant when the radius is 8 centimetres

30. A vessel in the form of an inverted cone has semi-vertical angle of 30° and contains water to a depth of x cm. Show that the volume V cm³ of water in the vessel is given by $V = \frac{1}{3}\pi x^3 \tan^2 30°$. If water is poured into the vessel at the rate of 132 cubic cm per minute, at what rate is the surface of the water rising when the depth is 9 centimetres?

31. Find the values of x for which the functions f and g defined by
$$f(x) = 2x^2 + 7x - 4 \text{ and } g(x) = x^3 - x^2 - 2x + 7$$
are increasing at the same rate.

32. If $\dfrac{dy}{dx} = 5x - \dfrac{2}{x^3}$ and if $y = 10$ when $x = 2$ express y in terms of x.

33. If $\dfrac{dy}{dx} = \dfrac{6}{x^2} - 8x$ and if $y = 21$ when $x = -1$ express y in terms of x.

34. Integrate with respect to x,

(i) $4 - \dfrac{5}{x^2} + 3x^{\frac{1}{2}}$ \quad (ii) $4x^3 + 1 - \dfrac{3}{x^3}$

Evaluate the following,

35. $\displaystyle\int \left(x^2 + \dfrac{1}{x^2}\right) dx$

36. $\displaystyle\int (x^2 + 3x - 5) \, dx$

37. $\displaystyle\int \dfrac{2x+1}{x^3} \, dx$

38. $\displaystyle\int \dfrac{1+\sqrt{x}}{x^2} \, dx$

39. $\displaystyle\int (3-x)^{\frac{1}{2}} \, dx$

40. $\displaystyle\int (4-3t)^{\frac{1}{3}} \, dt$

41. $\displaystyle\int (2+3t)^3 \, dt$

42. $\displaystyle\int \sin x \cos x \, dx$

43. If $f'(x) = 2(x^2 - 9)$ and if f has a maximum turning value of 16, find the function defined by $f(x)$.

44. If $\dfrac{dy}{dx} = x(x-3)^2$ and if $y = 2$ when $x = 1$ find y in terms of x. Show that the curve represented by y has a minimum turning point and one point of inflexion and hence sketch the graph of y in the closed interval $[-1, 4]$.

Evaluate the definite integrals,

45. $\displaystyle\int_{-1}^{2} (x-1)(x+2)^2 \, dx$

46. $\displaystyle\int_{1}^{3} \left(x + \dfrac{1}{x}\right)^2 dx$

47. $\displaystyle\int_{1}^{16} \dfrac{1}{\sqrt{x}} \, dx$

48. $\displaystyle\int_{0}^{\pi/4} \sin 2x \, dx$

49. $\displaystyle\int_{-\pi/2}^{\pi/2} (2 \cos 2x + 1) \, dx$

50. $\displaystyle\int_{1}^{4} p \, . \, dq$, where $pq^{1.5} = 1$

51. $\displaystyle\int_{1}^{9} F \, . \, ds$, where $F = \dfrac{(1-s^2)^2}{s^2}$

52. Sketch the curve $y = 3\sqrt{x}$ in the interval $[0, 16]$. Calculate,
   (i) the area bounded by the curve, the x-axis and the lines $x = 0$ and $x = 16$;
   (ii) the value of k so that the area bounded by the curve, the x-axis and the lines $x = 0$ and $x = k$ is $\frac{1}{8}$ of the area between the lines $x = 0$ and $x = 16$.

53. Find the volume obtained by revolving about the x-axis the finite area bounded by the lines $x + 2y - 4 = 0$, $x = 3$ and the coordinate axes.
   If this area is revolved round the y-axis, show that the volume of the solid generated is 9π units^3.

54. The area between the x-axis, the lines $x = 3$ and $x = 6$ and the part of the circle $x^2 + y^2 - 20x = 0$ above the x-axis is revolved once round the x-axis. Calculate the volume of the solid of revolution obtained.

55. Calculate the area bounded by the x-axis, the lines $x = -\dfrac{\pi}{4}$ and $x = \dfrac{\pi}{4}$ and the curves with equation,
   (i) $\cos x$ \quad (ii) $\cos 2x$

56. (i) Sketch the curve $y = \cos 2x$, for $0 \leq x \leq \pi$.
   (ii) Solve the equation $\cos 2x = -\frac{1}{2}$ for $0 \leq x \leq \pi$.
   (iii) Show that the area of the segment cut off on the curve by the straight line $y = -\frac{1}{2}$ is $\dfrac{\sqrt{3}}{2} - \dfrac{\pi}{6}$ square units.
   (Note: the segment is the area between the curve $y = \cos 2x$ and the line $y = -\frac{1}{2}$.)

ANSWERS

COORDINATE GEOMETRY

Page 12, Assignment 1.1

1. (i) 5 (ii) $\sqrt{101}$ (iii) 13 (iv) $\sqrt{170}$ (v) $3\sqrt{26}$
   (vi) $\sqrt{89}$ (vii) $5ak$.
2. $AB = AC = \sqrt{73}$. 3. 6 or 0. 4. 3 or -9.
5. (i) $(3, \frac{5}{2})$ (ii) $(-4, -\frac{1}{2})$ (iii) $(0, -\frac{1}{2})$ (iv) $(-4, -6)$
6. $(6, 0), (\frac{19}{2}, \frac{7}{2}), (\frac{11}{2}, \frac{1}{2})$. 7. $M(-1, -1), G(-1, 1)$.
8. $\frac{1}{2}$. 9. $(-\frac{7}{2}, \frac{11}{2}), (1, 1), (\frac{11}{2}, -\frac{7}{2})$.
10. $(1, 5)$. 11. $\sqrt{53}, \sqrt{20}, (\frac{1}{2}, 0)$. 12. $(3, 9)$.
13. $\frac{\sqrt{181}}{2}, \frac{\sqrt{205}}{2}$. 14. (i) $M(-1, -1), N(3, 2)$.

Page 15, Assignment 1.2

1. (i) A straight line 25 cm from the ground and parallel to the ground.
   (ii) The arc of a circle.
   (iii) A vertical straight line.
   (iv) A straight line inclined to the ground.
   (v) A circle of radius 5 cm and centre the point $(0, 0)$.
   (vi) $\{(3, -1)\}$.
   (vii) $\{(x, y): x = \pm 4, y \in R\}$.
   (viii) The region of the coordinate plane where $x > 0$, i.e. $\{(x, y): x, y \in R, x > 0\}$.
   (ix) The region of the coordinate plane where $x > 0$ and $y < 0$, but not including the coordinate axes, i.e. $\{(x, y): x > 0, y < 0, x, y \in R\}$.

2. (i)

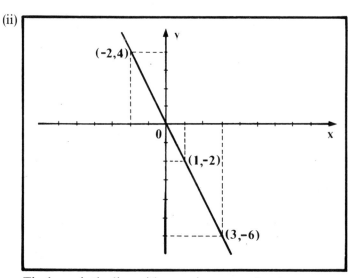

The locus is the line with equation $x = y$.

(ii)

The locus is the line with equation $y = -2x$.

(iii)

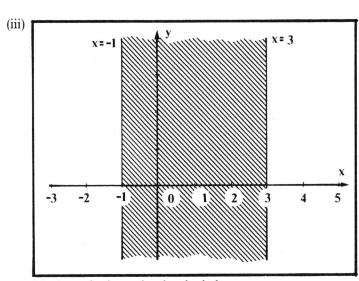

The locus is shown by the shaded area.
The boundaries are included.

(iv)

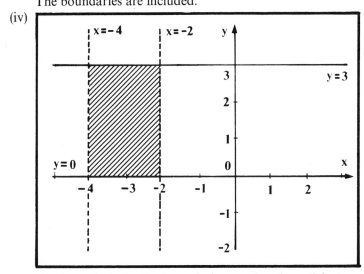

The locus includes the boundaries $y = 3$ and $y = 0$ but not $x = -4$ or $x = -2$

3

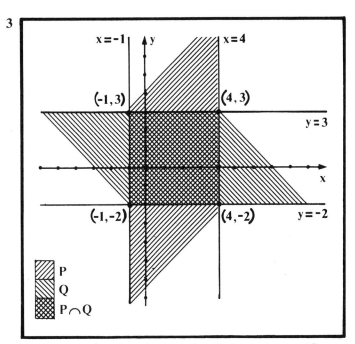

$P \cap Q = \{(x, y): -1 \leqq x \leqq 4, -2 \leqq y \leqq 3, x, y \in R\}$

4. $(-2, -4), (-1, -2), (0, 0), (1, 2), (2, 4)$.
5. On the line $x + 3y - 8 = 0$.   6. $x + y = 0$.
7. The circle with centre the point $(4, 2)$.
8. $PA = 2PM$; $x^2 - 3y^2 - 6x + 2y + 10 = 0$.
9. $x + 2y + 3 = 0$.
10. $3x^2 + 3y^2 - 16x - 20y + 32 = 0$; (iii), (iv).
11. (i) $\{(x, y): x = 5, y \in R\}$
    (ii) $\{(x, y): x^2 + y^2 - 16x + 60 = 0, x \in R, y \in R\}$
    (iii) $\{(x, y): x^2 + y^2 = 16, x \in R, y \in R\}$
    (iv) $\{(x, y): 2x^2 + 2y^2 - 35x + 143 = 0, x \in R, y \in R\}$

Locus of (i) is a straight line.
Locus of (ii) is a circle.

12. $y = 6$; locus is the straight line $y = 6$; $(a, 6)$, $(6, k)$ with $k = 6$.

Page 17, Assignment 1.3
1. A. 2. C. 3. E. 4. D. 5. D. 6. B.
7. B. 8. A. 9. D. 10. D. 11. C. 12. B.

Page 20, Assignment 2.1
1. (i) 30° (ii) 45° (iii) 153·45° (iv) 126·9°
   (v) 120° (vi) 153·45°.
2. (i) 1 (ii) $\frac{1}{4}$ (iii) $-\frac{5}{7}$ (iv) $\frac{7}{12}$ (v) $\frac{1}{2}$ (vi) $-\frac{8}{9}$
   (vii) $\frac{b}{a}$ (viii) $\frac{k}{h}$.
4. No. 5. (i), (ii), (iii). 6. -11.
7. $(1, 2), (-\frac{5}{2}, 3)$.
8. $(-2, 3), (-3, -1)$ and $(5, 1), (4, -3)$; $(-2, 3), (5, 1)$ and $(-3, -1), (4, -3)$.
9. $(-1, 4), (0, 6)$.

Page 23, Assignment 2.2
1. (i) $y = 2x$ (ii) $3y + x = 0$.
2. (i) $y = x$ (ii) $3y + 4x = 0$ (iii) $x = 0$.
3. (i) $-3, (0, 4)$ (ii) $\frac{1}{2}, (0, -1)$ (iii) $3, (0, -\frac{5}{2})$
   (iv) $3, (0, \frac{8}{3})$ (v) $-\frac{1}{3}, (0, 2)$ (vi) $-\frac{1}{2}, (0, 2)$
   (vii) $2, (0, -7)$ (viii) $-2, (0, \frac{4}{3})$.
4. (ii), (iii). 5. 2. 6. $aq = bp$.
7. (i) $y + 2x = 2$ (ii) $y = 3x - 7$ (iii) $2y + x = 10$.
8. See figure 23 opposite. 9. See figure 24 opposite.
10. (i) $2x - 3y = 7$ (ii) $4y + x + 18 = 0$
    (iii) $5y + 2x + 20 = 0$ (iv) $2x + 3y + 7 = 0$
    (v) $3x - 4y = 27$.
11. (i) $x = -2$ (ii) $y = 3$.

12. (i) $\{(4, -1)\}$ (ii) $\{(-3, -3)\}$.
13. (i) $y = 2x$ (ii) $2y = x$ (iii) $3y = x$ (iv) $4y = 3x$.
14. (i), (iii), (iv), (vii), (viii), (ix).
15. $y = 2x + 7$. 16. 3.
17. $(3, 2)$; they are concurrent.
18. $(4, 3), (1, 1), (-5, -3)$.

Page 29, Assignment 2.3
1. (i) $y = 3x - 9$ (ii) $y = -2x - 5$ (iii) $y = -4x - 11$
   (iv) $2y = x - 7$ (v) $3y = -2x + 4$
   (vi) $y = kx + q - kp$ (vii) $3y = -9x + 1$.
2. (i) $6y + 5x - 12 = 0$ (ii) $y + 4x = 13$
   (iii) $3y = 2x - 16$ (iv) $4y + x = -27$
   (v) $y + x = 3$ (vi) $2y + x = -14$.
3. (i) $x - 3y = 17$ (ii) $2x + y + 10 = 0$.
4. $y + 2x + 2 = 0$; $(-1, 0), (0, -2)$.
5. (i) $-\frac{1}{3}$ (ii) $\frac{1}{2}$ (iii) $-\frac{3}{2}$ (iv) 4 (v) -1 (vi) $\frac{5}{3}$
   (vii) $-\frac{2}{3}$.
6. (i) $-\frac{1}{2}$ (ii) $-\frac{3}{2}$ (iii) $\frac{1}{2}$ (iv) 2 (v) $-\frac{1}{6}$ (vi) -4.
7. (i) $2y + x = 0$ (ii) $y = x + 1$ (iii) $y = -2$
   (iv) $x = 3$.
8. $y + 3x = 12$. 9. $2y - x + 1 = 0$.
11. $3y - x + 3 = 0$. 12. 1.
13. $5y + x + 3 = 0, y + 2x - 2 = 0, 2y - 5x + 9 = 0$.
14. $y + x + 1 = 0, y - x + 3 = 0$; $(-1, -4)$.
15. (i) $(1, 1)$ (ii) $(3, 4)$
    (iii) no point of intersection; the lines are parallel
    (iv) $(-1, 1)$ (v) $(-3, -2)$.
17. $(-2, 2), (-4, -2), (4, -4)$. 18. $(-2, 2)$.
19. $(3, -2)$.

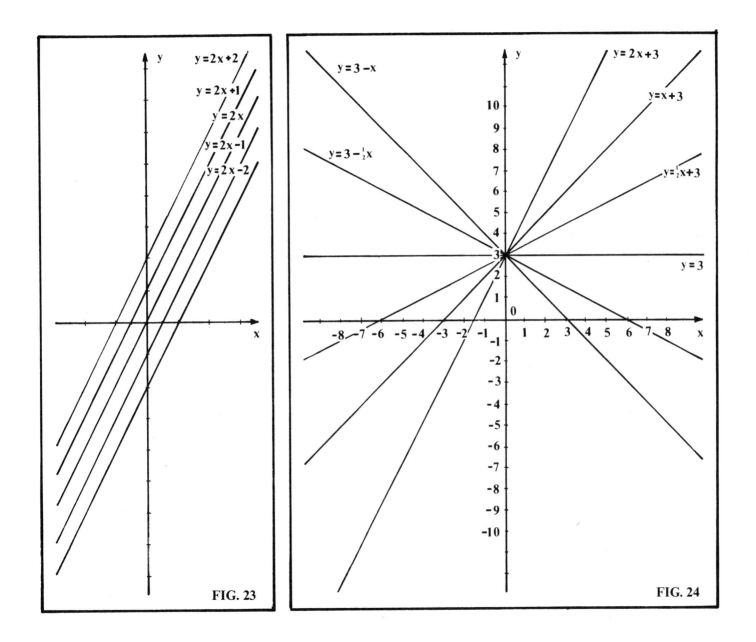

FIG. 23

FIG. 24

20. (i) $\sqrt{10}, 5\sqrt{2}$ (ii) $2y+x=4, y-2x=7$
(iii) $(-2,3);(-1,5)$.
21. $(-2,2)$. 22. $(\frac{11}{7}, \frac{4}{7})$. 23. $(-\frac{1}{3}, -1)$.

Page 31, Assignment 2.4
1. $(24, -16)$. 2. $(4, -1)$. 3. $x^2+y^2=16$.
4. $(-8, 23)$. 5. $(-3, 4); (-4, -6)$.
6. $y+x=6, 5y+x=26, y-3x=2$.
7. $y=2x+3$. 8. (i) $2:9$ (ii) $(10\frac{1}{2}, 5\frac{1}{2})$.
9. $4y-x=-3, 5y-3x=-9; 2:1$.
10. $3y-5x=18; (0, 6); 8\frac{1}{2}$ units^2.
11. $5y+3x=15$; (i) $(0, -\frac{25}{3})$; (ii) $\frac{85}{3}$ units^2.
12. $3x-4y=3$. 13. $(-3, 8)$.
14. (i) $A(\frac{15}{4}, \frac{9}{4})$, $B(0, 1)$, $C(\frac{5}{2}, -\frac{3}{2})$
16. (i) $(3, 0), (1, -1)$ (ii) $\sqrt{13}$.

Page 32, Assignment 2.5
1. B. 2. E. 3. E. 4. A. 5. E. 6. C.
7. C. 8. B. 9. B. 10. A, 11. A. 12. C.

Page 37, Assignment 3.1
1. (i) $x^2+y^2=4$ (ii) $x^2+y^2=12$ (iii) $4x^2+4y^2=9$
   (iv) $x^2+y^2=k^2$ (v) $x^2+y^2=20$ (vi) $x^2+y^2=18$
   (vii) $x^2+y^2=41$ (viii) $x^2+y^2=a^2+b^2$
   (ix) $x^2+y^2=a^2$.
2. $(4, -3)$ or $(-4, -3)$.
3. $(-2, 5), (5, 2), (-5, -2)$ lie on the circle
   $(3, -4), (1, -3)$ lie inside the circle
   $(0, -9), (4, -4), (-6, 1)$ lie outside the circle.
4. (i) A (ii) C (iii) B (iv) A (v) B (vi) B
   (vii) A (viii) C (ix) A.

5. (i) $(1, 0), (-1, 0), (0, 1), (0, -1)$
   (ii) $(3, 0), (-3, 0), (0, 3), (0, -3)$
   (iii) $(2\sqrt{3}, 0), (-2\sqrt{3}, 0), (0, 2\sqrt{3}), (0, -2\sqrt{3})$
   (iv) $(6\sqrt{2}, 0), (-6\sqrt{2}, 0), (0, 6\sqrt{2}), (0, -6\sqrt{2})$.
6. (i) $(3, 3), (-3, -3)$ (ii) $(4, -4), (-4, 4)$
   (iii) $(3, -1), (-1, 3)$ (iv) $(2, 5), (-5, -2)$.
7. $x^2+y^2=4$; a circle centre the origin and of radius 2 units.
8. $x^2+y^2=9$; a circle centre the origin and of radius 3 units.
9. $(1, -7), (1, 7), (-1, -7)$. 10. $x^2+y^2=2$.
11. $x^2+y^2=8$. 12. $x^2+y^2=36$.
13. $x^2+y^2=5$.

Page 39, Assignment 3.2
1. (i) $(x-1)^2+(y-2)^2=9$ (ii) $(x+1)^2+(y+2)^2=9$
   (iii) $x^2+(y-4)^2=25$ (iv) $(x+5)^2+(y-2)^2=36$
   (v) $(x+3)^2+(y+3)^2=16$ (vi) $(x-\frac{1}{2})^2+(y+\frac{1}{2})^2=4$.
2. (i) $(3, 2); 2$ (ii) $(-3, -1); 3$ (iii) $(2, -5); \sqrt{7}$
   (iv) $(-5, 4); 2\sqrt{6}$.
3. (i) outside (ii) inside (iii) outside (iv) on
   (v) on (vi) inside (vii) on.
4. 7. 5. $3\sqrt{5}$. 6. $x^2+y^2-4x-14y+4=0$.
7. $x^2+y^2+6x-6y+9=0$.
8. (i) $(x-5)^2+(y+2)^2=9; x^2+y^2-10x+4y+20=0$
   (ii) $(x+2)^2+(y+1)^2=81; x^2+y^2+4x+2y-76=0$
   (iii) $(x-3)^2+(y+4)^2=25; x^2+y^2-6x+8y=0$
   (iv) $(x+3)^2+(y+2)^2=41; x^2+y^2+6x+4y-28=0$
   (v) $(x-2)^2+(y-2)^2=41; x^2+y^2-4x-4y-33=0$.
9. (i) $x^2+y^2-4x-6y+4=0$

(ii) $x^2 + y^2 + 10x + 10y + 37 = 0$
(iii) $x^2 + y^2 - 6x + 4y = 0$.
10. $y + 2x = 5$; $(5, -5)$.
11. $x^2 + y^2 + 2x - 10y + 25 = 0$;
   (i) $x^2 + y^2 - 2x - 10y + 25 = 0$
   (ii) $x^2 + y^2 + 2x + 10y + 25 = 0$
   (iii) $x^2 + y^2 - 2x + 10y + 25 = 0$.

Page 42, Assignment 3.3

1. (i), (ii), (iii), (vi), (vii), (viii).
2. (i) $(-1, -4)$; 5 (ii) $(3, -3)$; 4 (iii) $(-2, -5)$; 4
   (iv) $(-3, -4)$; 5 (v) $(\frac{5}{2}, \frac{9}{2})$; $\frac{7\sqrt{2}}{2}$ (vi) $(-\frac{7}{4}, \frac{3}{4})$; $\frac{\sqrt{66}}{4}$
   (vii) $(\frac{3}{2}, -2)$; 3 (viii) $(-\cos\theta, -\sin\theta)$; 1
   (ix) $(-2a, -2a)$; $3a$.
3. (iii)   4. -4 or 1.   5. $(6, 0), (-2, 0)$.
6. $x^2 + y^2 + 2x - 6y - 15 = 0$.
9. $x^2 + y^2 - 4x - 3y = 0$.
10. (i) $x^2 + y^2 - 5x - 3y = 0$ (ii) $x^2 + y^2 - 5x + y + 4 = 0$
    (iii) $x^2 + y^2 + 7x - y + 10 = 0$
    (iv) $x^2 + y^2 + 6x + 2y + 5 = 0$.
11. (i) $x^2 + y^2 + 4x - 2y = 0$
    (ii) $x^2 + y^2 - 2x - 6y - 15 = 0$
    (iii) $x^2 + y^2 - 5x - 12y + 41 = 0$
    (iv) $x^2 + y^2 - 2x - 3y - 1 = 0$.
12. $x^2 + y^2 + y - 4 = 0$.   13. $x^2 + y^2 - 3x - 20 = 0$.
14. $3x^2 + 3y^2 + 16y + 16 = 0$.

Page 46, Assignment 3.4

1. (i) $(3, 1), (-1, -3)$ (ii) $(-1, -1), (\frac{7}{5}, \frac{1}{5})$
   (iii) $(-2, -1), (\frac{11}{5}, \frac{2}{5})$ (iv) $(-2, -1), (-1, 0)$

(v) $(3, 2), (1, -2)$ (vi) $(4, 1), (1, 2)$.
2. (i) $(3, -3)$ (ii) $(3, -3)$ (iii) $(0, 2)$ (iv) $(4, -6)$
   (v) $(1, 1)$.
3. (i) $y = 2x - 5, y = -2x - 5$
   (ii) $2y = x + 5, 2y = -x - 5$
   (iii) $2y = \sqrt{5}x + 6, 2y = -\sqrt{5}x + 6$
   (iv) $y = x - 4, y = -x - 4$
   (v) $\sqrt{2}y = x - 6, \sqrt{2}y = -x + 6$.
4. (i) $3y - 2x + 13 = 0$ (ii) $5y + x + 26 = 0$
   (iii) $y - 2x - 10 = 0$ (iv) $x + 2 = 0$
   (v) $y + x + 8 = 0$ (vi) $x = 3$ (vii) $y + x = 0$.
5. (i) 4 (ii) 12 (iii) $2\sqrt{21}$ (iv) $5\sqrt{2}$ (v) $\sqrt{57}$
   (vi) $2\sqrt{13}$.
7. $(4, -2)$; $-\frac{11}{2}$.   8. $y = 3$; $x = -3$.
9. $(2, -4)$.   10. ± 10.   11. $y + x = 3, y - x = 1$; $(1, 2)$.
12. $(0, 6), (\frac{44}{5}, \frac{8}{5})$; $3y - 4x - 18 = 0, 7y - 24x + 200 = 0$.

Page 47, Assignment 3.5

1. $\pm\frac{4}{3}$; $(-\frac{12}{5}, \frac{9}{5}), (\frac{12}{5}, \frac{9}{5})$.
2. $x^2 + y^2 + 10x - 10y + 25 = 0$ and
   $x^2 + y^2 - 6x - 10y + 9 = 0$.
3. $x^2 + y^2 + 2x - 1 = 0$.
4. $x^2 + y^2 - 5x = 0$; $(\frac{5}{2}, 0)$, $\frac{5}{2}$.
5. $x^2 + y^2 + 6x - 8y = 0, (0, 0), (0, 8), (-6, 0)$;
   $4y + 3x + 18 = 0, 4y + 3x - 32 = 0, 4y - 3x = 0$;
   $(-3, -\frac{9}{4}), (\frac{16}{3}, 4), \frac{15}{4}, \frac{20}{3}$.
6. $(-1, 3)$; $\sqrt{5}$.
7. $(3, 0), 10$; (i) $(-7, 0), (9, 8)$; $(1, 4)$
   (ii) $x^2 + y^2 + 2x - 16y - 35 = 0$.
8. (i) $-1 < k < \frac{1}{7}$ (ii) -1 or $\frac{1}{7}$.

9. $(1,2), (\frac{2}{5}, -\frac{11}{5})$; (i) $y = 7x+5\sqrt{10}, y = 7x-5\sqrt{10}$
   (ii) $2\sqrt{5}$.
10. $(-3,4), 10; (3,-4), (-9,12)$.
11. (i) $2x^2+2y^2-15x-5y-20 = 0$
    (ii) $x^2+y^2-6x-3y-10 = 0$.
12. (i) $x^2+y^2-2x-2y+1 = 0$,
    $x^2+y^2-10x-10y+25 = 0$ (ii) $(2, 1)$ (iii) $(\frac{3}{2}, \frac{3}{2})$.
13. Roots give the gradients of the two tangents from the origin to the circle.
14. $(-4, -3), (8, 1)$. 15. $(\frac{1}{5}, \frac{2}{5})$. 16. $(-3, 2)$.
17. (i) $(-4, -2)$ (ii) $3x+4y = 0$.

Page 49, Assignment 3.6
1. D.   2. E.   3. E.   4. D.   5. E.   6. D.
7. A.   8. B.   9. E.   10. D.   11. C.   12. C.

TRIGONOMETRY

Page 57, Assignment 1.1
1. (i) − (ii) − (iii) − (iv) − (v) + (vi) −
   (vii) + (viii) + (ix) + (x) + (xi) − (xii) +.
2. (a) −, −, +   (b) +, +, +   (c) −, +, −
   (d) +, −, −.
3. $0 \leq x < 90, 270 < x < 360$.
4. $180 < x < 360$.
5. (i) 1st and 4th quadrants (ii) 4th quadrant
   (iii) 3rd quadrant (iv) 3rd quadrant.
6. (i) 0·616 (ii) −7·115 (iii) −0·956 (iv) −0·985
   (v) 0·695 (vi) −5·671 (vii) −1 (viii) 0.
7. (i) $\frac{\sqrt{3}}{2}$ (ii) $\frac{1}{\sqrt{2}}$ (iii) $\sqrt{3}$ (iv) $\frac{1}{\sqrt{2}}$ (v) −1
   (vi) −1 (vii) $-\frac{\sqrt{3}}{2}$ (viii) $\frac{1}{\sqrt{3}}$.
8. (i) {45} (ii) {120} (iii) {60, 120} (iv) {30, 150}
   v) ∅ (vi) {120} (vii) {109·8}.
9. (i) −0·951 (ii) −0·754 (iii) 0·454
   (iv) −0·707, $-\frac{1}{\sqrt{2}}$ (v) 0·866, $\frac{\sqrt{3}}{2}$
   (vi) −0·577, $-\frac{1}{\sqrt{3}}$.
10. (i) $-\cos a°$ (ii) $-\tan p°$ (iii) $\cos x°$ (iv) $\sin q°$
    (v) $\sin t°$.
11. (i) $-\cos 10°$ (ii) $\sin 62°$ (iii) $-\tan 88°$.
12. 45°. 13. $\frac{4}{5}$. 14. $\frac{5}{13}, -\frac{12}{13}$.
15. (i) ∅ (ii) {110} (iii) {148·4} (iv) {135}.
17. $-\frac{1}{2}, \frac{1}{2}, \frac{1}{2}$.

Page 60, Assignment 1.2
1. (i) −0·342 (ii) $-\frac{1}{\sqrt{2}}$ (iii) $-\sqrt{3}$ (iv) −0·259
   (v) 1·192 (vi) $\frac{1}{\sqrt{2}}$ (vii) 0·643 (viii) $-\frac{1}{2}$
   (ix) 0·755 (x) −0·966 (xi) −1 (xii) 1.
2. (i) $\tan 35°$ (ii) $-\sin 14°$ (iii) $\cos 88°$
   (iv) $-\sin 12°$ (v) $-\cos 89°$ (vi) $-\tan 50°$.
3. (i) $-\sin 50°$ (ii) $-\sin 60°$ (iii) $-\tan 20°$
   (iv) $\cos 60°$ (v) $-\cos 62°$ (vi) $-\tan 30°$.
4. (i) $-\frac{1}{\sqrt{2}}$ (ii) $\frac{\sqrt{3}}{2}$ (iii) −0·900 (iv) 0·342
   (v) $-\sqrt{3}$ (vi) 0·985 (vii) 1 (viii) −1 (ix) $\frac{1}{\sqrt{2}}$.
5. $-\frac{20}{29}, -\frac{21}{29}$ 6. $-\frac{5}{13}, -\frac{5}{12}$.
7. (i) {60, 300} (ii) {30, 210} (iii) {135, 225}

(iv) $\{210, 330\}$ (v) $\{135, 315\}$
9. (i) $\{229\cdot4, 310\cdot6\}$ (ii) $\{135, 315\}$ (iii) $\{26\cdot5, 333\cdot5\}$
   (iv) $\{51\cdot4, 231\cdot4\}$.

Page 63, Assignment 1.3

1. (i) $\tan 25°$ (ii) $\tan 50°$ (iii) $\sin A$ (iv) $\sin A$
2. (i) 1 (ii) 1.   3. 1.   4. $-\frac{1}{7}$.
6. (i) $4 - 3\cos^2 A$ (ii) $4\sin^2 A - 3$.
14. (i) $x^2 + y^2 = 16$ (ii) $\frac{x^2}{a^2} + \frac{y^2}{b^2} = 1$ (iii) $m^2 + n^2 = 5$
    (iv) $a^2 + b^2 = 2$.
15. $(2-y)(1-xy) = 0$.

Page 67, Assignment 1.4

1. (i) Max. value 1 at $x = 90$, min. value -1 at $x = 270$
   (ii) Max. value 1 at $x = 0$ or 360, min. value -1 at $x = 180$
   (iii) No max. or min. values.
2. (i) 36, 144 (ii) 53, 307 (iii) 129, 309.
3. $x = 97$; max. value $= 2\cdot2$, when $x = 63$.
5. (i) $\{135\}$ (ii) $\{53\}$ (iii) $\{90, 180\}$.
6. (i) $2\cdot2, 116°$ (ii) $52 \leq t \leq 90$ and $143 \leq t \leq 180$
   (iii) $a = 2, b = 1$.
7. (i) $\{0, 90, 180, 270\}$ (ii) $p = 3, k = 2$
   (iii) $22\cdot5 \leq x \leq 67\cdot5, 202\cdot5 \leq x \leq 247\cdot5$
   (iv) $90 \leq x \leq 112\cdot5, 157\cdot5 \leq x \leq 180$.
8. (i) $360°, 180°, 180°$ (ii) $180°, 90°, 180°$.
9. (i) $360°$ (ii) $180°$ (iii) $180°$ (iv) $180°$.
10. (i) $120°$ (ii) $720°$ (iii) $360°$ (iv) $90°$.

Page 72, Assignment 1.5

2. (i) $(4, -330°), (-4, 210°), (-4, -150°)$
   (ii) $(2, -60°), (2, 300°), (-2, -240°)$
   (iii) $(1, 130°), (-1, 310°), (1, -230°)$
   (iv) $(-3, 120°), (-3, -240°), (3, 300°)$
   (v) $(3, -60°), (-3, 120°), (3, 300°)$
   (vi) $(3, 60°), (-3, 240°), (-3, -120°)$.
3. (i) $(3, 30°)$ (ii) $(2, 150°)$ (iii) $(1, 220°)$ (iv) $(2, 310°)$
   (v) $(2, 170°)$.
4. (i) $(2, 90°)$ (ii) $(3, 0°)$ (iii) $(2, 30°)$ (iv) $(2, 300°)$
   (v) $(2, 315°)$ (vi) $(\sqrt{2}, 225°)$ (vii) $(13, 67\cdot4°)$
   (viii) $(10, 126\cdot9°)$ (ix) $(3\sqrt{5}, 333\cdot45°)$.
5. (i) $(0, 4)$ (ii) $(-2, 0)$ (iii) $(-1, 1)$ (iv) $\left(\frac{3}{2}, -\frac{\sqrt{3}}{2}\right)$
   (v) $\left(-\frac{\sqrt{3}}{2}, -\frac{3}{2}\right)$ (vi) $(\sqrt{2}, -\sqrt{2})$ (vii) $(3, 0)$
   (viii) $(0, -8)$ (ix) $(-3\cdot76, -1\cdot37)$.

Page 76, Assignment 1.6

1. $6\cdot12$. 2. $94\cdot6°$. 3. $41\cdot7°$; $77\cdot8$ units^2.
4. $37\cdot5°, 142\cdot5°$. 5. $\frac{1}{2}pq \sin R$; $18\cdot2$ cm. 6. $16\cdot4$ km.
7. $78\cdot25°$. 9. (ii) $\frac{d \cos \beta}{\sin(\beta - \alpha)}$. 12. $3\cdot23$ km, $2\cdot91$ km.
13. 175 km, $138\cdot1°$.
15. (i) $\sqrt{7}$ (ii) $\sqrt{13}$ (iii) $3\sqrt{3}$ (iv) 5.
16. $4\cdot98, \sqrt{7}$.

Page 79, Assignment 1.7

1. C 2. B 3. A 4. B 5. B 6. D 7. D 8. D
9. A 10. E 11. A 12. B.

Page 84, Assignment 2.1

1.
angle in degrees	0	30	60	180	120	150
angle in radians	0	$\frac{\pi}{6}$	$\frac{\pi}{3}$	π	$\frac{2\pi}{3}$	$\frac{5\pi}{6}$

angle in degrees	45	90	135	180	360
angle in radians	$\frac{\pi}{4}$	$\frac{\pi}{2}$	$\frac{3\pi}{4}$	π	2π

2. $\frac{2\pi}{9}, \frac{9\pi}{20}, \frac{3\pi}{5}, \frac{5\pi}{4}, 3\pi$.

3. $22\cdot5°, 114\cdot6°, 100°, 85\cdot9°, 194\cdot8°$.

4. (i) $\frac{1}{\sqrt{2}}$ (ii) $-\frac{1}{2}$ (iii) 0 (iv) $-\frac{1}{\sqrt{3}}$ (v) $-\frac{\sqrt{3}}{2}$ (vi) $-\frac{\sqrt{3}}{2}$.

5. $0\cdot75, 1\cdot26, 1\cdot92, 6\cdot98$. 6. $25\cdot8°, 45\cdot8°, 80\cdot2°, 143\cdot8°$.

7. (i) $\sin\theta$ (ii) $-\tan\theta$ (iii) $-\sin\theta$ (iv) $-\sin\theta$ (v) $\cos\theta$.

8. (i) $\left\{\frac{\pi}{3}, \frac{2\pi}{3}\right\}$ (ii) $\{0, 2\pi\}$ (iii) $\left\{\frac{\pi}{4}, \frac{5\pi}{4}\right\}$ (iv) $\{0, \pi, 2\pi\}$ (v) $\left\{\frac{2\pi}{3}, \frac{5\pi}{3}\right\}$.

9. (i) $\{0, 2\pi\}$ (ii) $\left\{\frac{\pi}{3}, \frac{5\pi}{3}\right\}$ (iii) $\{0, \pi, 2\pi\}$.

10. $\frac{\pi r x}{180}$. 12. (i) $1\cdot6$ (ii) $91\cdot7°$.

Page 86, Assignment 2.2

1. D 2. B 3. D 4. B 5. D 6. E 7. B 8. C 9. E 10. A 11. D 12. B.

Page 92, Assignment 2.3

1. (i) $\sin x \cos y + \cos x \sin y$
   (ii) $\cos a° \cos b° - \sin a° \sin b°$
   (iii) $\cos\theta \cos 2\alpha + \sin\theta \sin 2\alpha$
   (iv) $\sin 2\theta° \cos \beta° - \cos 2\theta° \sin \beta°$.

3. (i) $\sin 55°$ (ii) $\cos 94°$ (iii) $\cos(2x - y)$ (iv) $\sin 2\theta$ (v) $\sin 3\phi$ (vi) $\cos 4\beta$

4. (i) $\frac{\sqrt{3}+1}{2\sqrt{2}}$ (ii) $\frac{\sqrt{3}-1}{2\sqrt{2}}$. 10. (i) $\frac{\sqrt{3}}{2}$ (ii) 1.

11. (i) $\frac{1}{2}$ (ii) 0 (iii) 0. 12. 1.

13. (i) $\frac{1+\sqrt{3}}{2\sqrt{2}}$ (ii) $\frac{\sqrt{3}-1}{2\sqrt{2}}$. 14. $-\frac{16}{65}, \frac{63}{65}$.

15. (i) $\{30\}$ (ii) $\{30\}$ (iii) $\left\{\frac{\pi}{6}, \frac{5\pi}{6}\right\}$ (iv) $\left\{\frac{\pi}{6}, \frac{5\pi}{6}\right\}$.

16. $\{49\cdot1, 229\cdot1\}$. 19. $\frac{1}{\sqrt{2}}$. 20. $-\frac{1}{\sqrt{5}}$.

Page 94, Assignment 2.4

1. (i) $\frac{\tan x + \tan y}{1 - \tan x \tan y}$ (ii) $\frac{\tan 2\alpha + \tan \beta}{1 - \tan 2\alpha \tan \beta}$ (iii) $\frac{\tan 3\theta - \tan \phi}{1 + \tan 3\theta \tan \phi}$.

3. (i) $\frac{1}{\sqrt{3}}$ (ii) $\sqrt{3}$. 4. $2-\sqrt{3}, -(2+\sqrt{3})$. 7. $\frac{4}{33}$.

10. (i) $\{45, 116\cdot6, 225, 296\cdot6\}$ (ii) $\{0, 90, 180, 270, 360\}$.

12. $\frac{284}{257}$.

Page 97, Assignment 2.5

1. $2\sin\theta\cos\theta$, $\cos^2 A - \sin^2 A$ or $2\cos^2 A - 1$ or $1 - 2\sin^2 A$, $\frac{2\tan B}{1 - \tan^2 B}$.

2. $2\sin 2\theta \cos 2\theta$, $\cos^2 2x - \sin^2 2x$ or $2\cos^2 2x - 1$ or $1 - 2\sin^2 2x$, $\frac{2\tan 2\phi}{1 - \tan^2 2\phi}$.

3. $\cos^2 \tfrac{1}{2}\theta - \sin^2 \tfrac{1}{2}\theta$, $2\cos^2 \tfrac{1}{2}\theta - 1$, $1 - 2\sin^2 \tfrac{1}{2}\theta$.

4. (i) $\dfrac{24}{25}, \dfrac{7}{25}, \dfrac{24}{7}$ (ii) $\dfrac{\sqrt{3}}{2}, -\dfrac{1}{2}, -\sqrt{3}$
 (iii) $\dfrac{120}{169}, \dfrac{119}{169}, \dfrac{120}{119}$.

5. (i) $\dfrac{\sqrt{3}}{2}, -\dfrac{1}{2}, -\sqrt{3}$ (ii) $\dfrac{240}{289}, \dfrac{161}{289}, \dfrac{240}{161}$
 (iii) -1, 0, no finite value.

6. (i) 1 (ii) $\tan 2\alpha$ (iii) $\sin 4\theta$ (iv) $\cos 36°$
 (v) $\cos 4x$ (vi) $2\cos^2 A$ (vii) $\tan \alpha$.

7. (i) $\tfrac{1}{2}$ (ii) $-\tfrac{1}{2}$ (iii) $\sqrt{3}$ (iv) $\dfrac{\sqrt{3}}{2}$. 9. 1.

10. $\tfrac{1}{2} \sin 2\theta$; $\dfrac{\pi}{4}, \dfrac{5\pi}{4}$; $\tfrac{1}{2}$. 11. $p = 8$, $q = -8$, $r = 1$.

12. $a = \tfrac{1}{8}$, $b = \tfrac{1}{2}$, $c = \tfrac{3}{8}$.

18. (i) Max. value $= \tfrac{11}{3}$, min. value $= -7$
 (ii) Max. value $= 1\tfrac{1}{8}$, min. value $= -2$
 (iii) Max. value $= 7$, min. value $= -\tfrac{11}{3}$.

26. $RQ' = 5$, $PQ' = 10$; $\dfrac{\pi}{2} - 2\alpha$.

27. (i) $\{15, 75\}$ (ii) $\{75, 105\}$ (iii) $\left\{\dfrac{\pi}{9}, \dfrac{4\pi}{9}, \dfrac{7\pi}{9}\right\}$.

28. (i) $\{30, 90, 150, 270\}$ (ii) $\{0, 120, 180, 240\}$
 (iii) $\left\{0, \dfrac{\pi}{2}, \pi, \dfrac{3\pi}{2}, 2\pi\right\}$.

29. $\left\{0, \dfrac{2\pi}{3}, \dfrac{4\pi}{3}, 2\pi\right\}$. 30. $\left\{0, \pi, \dfrac{7\pi}{6}, \dfrac{11\pi}{6}, 2\pi\right\}$.

31. (i) $\{0, 60, 120, 180, 240, 300, 360\}$ (ii) $\{90, 228.6, 311.4\}$
 (iii) $\left\{\dfrac{\pi}{8}, \dfrac{5\pi}{8}, \dfrac{9\pi}{8}, \dfrac{13\pi}{8}\right\}$.

32. $\{90\}$.

33. (i) $\{0, 73.2, 106.8, 180\}$ (ii) $\{45, 97.25, 135, 172.75\}$.

34. $\{0, 74.35, 180, 285.65, 360\}$. 35. $\{199.45, 340.55\}$.

Page 99, Assignment 2.6

1. B 2. B 3. D 4. A 5. E 6. B 7. C 8. B
9. E 10. E 11. B 12. C.

Page 103, Assignment 3.1A

1. $2 \sin 19° \cos 9°$. 2. $2 \cos 25° \cos 12°$.
3. $2 \cos 21° \sin 8°$. 4. $2 \sin 30° \sin 12°$.
5. $2 \cos 78° \sin 27°$. 6. $-2 \sin 38° \sin 18°$.
7. $2 \cos 34° \cos 5°$. 8. $2 \cos 350° \sin 60°$.
9. $2 \cos 2x° \cos x°$. 10. $2 \sin(\alpha - \beta) \sin \beta$.
11. $2 \cos\left(x + \dfrac{y}{2}\right) \sin \dfrac{y}{2}$. 12. $-2 \sin\left(x + \dfrac{h}{2}\right) \sin \dfrac{h}{2}$.
13. $2 \cos x \cos \dfrac{\pi}{2}$. 14. $2 \sin 125° \cos 56°$.
15. $2 \cos\left(x + \dfrac{h}{2}\right) \sin \dfrac{h}{2}$. 16. $-2 \sin 2a° \sin a°$.
17. $2 \cos \tfrac{1}{2}\phi \sin \tfrac{1}{6}\phi$. 18. $2 \sin \alpha \cos \beta$.
19. $2 \sin(2x - 11)° \sin(x - 1)°$.
20. $-2 \cos\left(\dfrac{\pi}{4} - \dfrac{3}{2}\theta\right) \sin \dfrac{\theta}{2}$.
21. $2 \cos \dfrac{\pi}{4} \cos \dfrac{\pi}{8}$. 22. $2 \cos 3\theta \cos 2\theta$.
23. $2 \cos 5A \sin 3A$. 24. $2 \cos 10° \cos 5°$.
25. $\dfrac{1}{\sqrt{2}}$. 26. 0. 27. $\dfrac{1}{\sqrt{2}}$. 28. -1. 29. $\dfrac{1}{\sqrt{3}}$.

Page 104, Assignment 3.1B

1. $\sin 66° + \sin 26°$. 2. $\sin 44° - \sin 20°$.

3. $\cos 60° + \cos 18°$.
4. $\cos 7° - \cos 65°$.
5. $\sin(\alpha+\beta) + \sin(\alpha-\beta)$.
6. $\cos(x-y) - \cos(x+y)$.
7. $\cos(A+B) + \cos(A-B)$.
8. $\sin(Q+P) - \sin(Q-P)$.
9. $\tfrac{1}{2}(\sin 55° + \sin 33°)$.
10. $\tfrac{1}{2}(\sin 55° - \sin 33°)$.
11. $\cos 2x + \cos 2y$.
12. $\sin \dfrac{7\pi}{8} + \sin \dfrac{3\pi}{8}$.
13. $\sin 2\theta + \sin \theta$.
14. $\cos 2\beta + \cos \dfrac{2\pi}{3}$.
15. $\sin 2A + \sin 2B$.
16. $\sin 10\theta + \sin 4\theta$.
17. $\cos 2\beta - \cos 4\alpha$.
18. $\sin \alpha + \sin \beta$.
19. $\cos \beta - \cos \alpha$.
20. $\cos 3P + \cos(P+2Q)$.
21. $\sin 48° + \sin 2\alpha°$.
22. $\sin 2x - \sin(2y-2z)$.
26. $\cos(P+2Q) - \cos(3P+8Q)$.
27. $\sin(4\alpha+6\beta) + \sin(2\alpha+2\beta)$.
28. $\sin(6\alpha+2\beta) - \sin(4\alpha+4\beta)$.
30. (i) Max. value = 2, Min. value = 0
    (ii) Max. value = 1·643, Min. value = −0·357
    (iii) Max. value = 0·234, Min. value = −1·766.
32. Max. value for $\theta = 15$ or 195, Min. value for $\theta = 105$ or 285.

Page 107, Assignment 3.2

1. (i) $\{60, 120\}$
   (ii) $\{60 + n.360\} \cup \{120 + n.360\}, n \in Z$.
2. $\{60 + n.360\} \cup \{300 + n.360\}, n \in Z$.
3. (i) $\{210, 330\}$
   (ii) $\{210 + n.360\} \cup \{330 + n.360\}, n \in Z$.
4. $\{x : x = 45 + n.180, n \in Z\}$.
5. (i) $\left\{\dfrac{\pi}{4}, \dfrac{3\pi}{4}\right\}$ (ii) $\left\{\dfrac{\pi}{4} + 2n\pi\right\} \cup \left\{\dfrac{3\pi}{4} + 2n\pi\right\}, n \in Z$.
6. $\left\{x : x = \dfrac{2\pi}{3} + n\pi, n \in Z\right\}$.

7. $\{x : x = 64\cdot3 + n.180, n \in Z\}$.
8. $\{143\cdot5 + n.360\} \cup \{216\cdot5 + n.360\}, n \in Z$.
9. $\left\{\alpha : \alpha = \dfrac{\pi}{2} + 2n\pi, n \in Z\right\}$.
10. $\{\alpha : \alpha = \pi + 2n\pi, n \in Z\}$.
11. (i) $\{0, \pi\}$ (ii) $\{\alpha : \alpha = n\pi, n \in Z\}$.
12. $\{45, 135, 225, 315\}$. 13. $\{20, 40, 140, 160, 260, 280\}$.
14. $\{45, 105, 165, 225, 285, 345\}$.
15. $\left\{0, \dfrac{\pi}{2}, \pi, \dfrac{3\pi}{2}, 2\pi\right\}$. 16. $\left\{\dfrac{\pi}{4}, \dfrac{5\pi}{4}\right\}$.
17. $\left\{\dfrac{\pi}{4}, \dfrac{5\pi}{12}, \dfrac{11\pi}{12}, \dfrac{13\pi}{12}, \dfrac{19\pi}{12}, \dfrac{21\pi}{12}\right\}$. 18. $\left\{\dfrac{\pi}{3}, \dfrac{5\pi}{3}\right\}$.
19. $\left\{\dfrac{5\pi}{3}\right\}$. 20. $\{0, 45, 135, 180, 225, 315, 360\}$.
21. $\{0, 90, 180, 270, 360\}$. 22. $\left\{\dfrac{\pi}{3}, \dfrac{\pi}{2}, \dfrac{2\pi}{3}, \dfrac{4\pi}{3}, \dfrac{3\pi}{2}, \dfrac{5\pi}{3}\right\}$.
23. $\{0, 45, 60, 135, 180, 225, 300, 315, 360\}$.
24. $\left\{0, \dfrac{3\pi}{8}, \dfrac{7\pi}{8}, \pi, \dfrac{11\pi}{8}, \dfrac{15\pi}{8}, 2\pi\right\}$.
25. $\left\{0, \dfrac{\pi}{2}, \pi, \dfrac{7\pi}{6}, \dfrac{3\pi}{2}, \dfrac{11\pi}{6}, 2\pi\right\}$.
26. $\{15, 75, 90, 195, 255, 270\}$. 27. $\{0, 120, 240, 360\}$.
28. $\{36, 90, 108, 180, 252, 270, 324\}$. 30. $\left\{x : 0 \leqq x < \dfrac{\pi}{3}\right\}$.

Page 108, Assignment 3.3

1. D 2. D 3. C 4. A 5. B 6. B 7. A 8. C
9. E 10. C 11. B 12. B.

Page 112, Assignment 4.1

1. (i) 5, 36·9 (ii) 13, 292·6 (iii) 17, 329 (iv) 2, 30
   (v) 10, 233·1 (vi) $2\sqrt{10}$, 108·4.

2. (i) $5\cos(x-53\cdot1)°$ (ii) $\sqrt{5}\cos(x-333\cdot45)°$
   (iii) $2\sqrt{5}\cos(x-153\cdot45)°$.
3. (i) $10\sin(x-216\cdot9)°$ (ii) $\sqrt{2}\sin(x-315)°$.
4. (i) $2\cos\left(\theta-\dfrac{\pi}{6}\right)$ (ii) $\sqrt{3}\sin\left(\theta-\dfrac{\pi}{180}\times 215\cdot2\right)$
   (iii) $\sqrt{2}\sin\left(\theta-\dfrac{3\pi}{4}\right)$.
5. (i) $2\sqrt{2}\cos(x-120)°$ (ii) $\sqrt{13}\cos(x-33\cdot7)°$
   (iii) $2\cdot6\cos(x-292\cdot6)°$ (iv) $\sqrt{2}\cos\left(x-\dfrac{7\pi}{4}\right)$
   (v) $2\cos\left(x-\dfrac{\pi}{3}\right)$.

Page 113, Assignment 4.2

1. (i) Max. 3 at $x=90$, Min. -3 at $x=270$
   (ii) Max. 1 at $x=27$, Min. -1 at $x=207$
   (iii) Max. 5 at $x=150$, Min. -5 at $x=330$
   (iv) Max. 1 at $x=45$ and 225, min. -1 at $x=135$ and 315
   (v) Max. 3 at $x=5$ and 185, min. -3 at $x=95$ and 275.
2. (i) Max. 10 at $x=36\cdot9$, min. -10 at $x=216\cdot9$
   (ii) Max. $\sqrt{2}$ at $x=315$, min. $-\sqrt{2}$ at $x=135$
   (iii) Max. 17 at $x=241\cdot9$, min. -17 at $x=62\cdot9$.
3. (i) Max. 2 at $\theta=\dfrac{3\pi}{4}$, min. -2 at $\theta=\dfrac{7\pi}{4}$
   (ii) Max. 2 at $\theta=\dfrac{\pi}{3}$, min. -2 at $\theta=\dfrac{4\pi}{3}$.
5. Max. 6, min. 4.
6. High at 0227, and 1427 hours, low at 0527 and 1727 hours.
7. $2d\sqrt{2}$.

Page 115, Assignment 4.3

1. $\{96\cdot9, 336\cdot9\}$. 2. $\{79, 235\cdot8\}$. 3. $\{233\cdot1, 323\cdot1\}$.
4. $\{96\cdot9, 336\cdot9\}$. 5. $\{180, 270\}$. 6. $\{154\cdot5, 303\cdot6\}$.
7. $\{62\cdot8, 170\cdot3, 242\cdot8, 350\cdot3\}$.
8. $\{16\cdot8, 126\cdot3, 196\cdot8, 306\cdot3\}$.
9. $\left\{\pi, \dfrac{3\pi}{2}\right\}$. 10. $\left\{\dfrac{2\pi}{3}\right\}$. 11. $\left\{0, \dfrac{5\pi}{6}, \pi, \dfrac{11\pi}{6}\right\}$.
12. Max. $7\sqrt{2}$, min. $-7\sqrt{2}$; $\{96\cdot8, 233\cdot2\}$.

Page 116, Assignment 4.4

1. E 2. D 3. C 4. D 5. C 6. C 7. B 8. D 9. E
10. E 11. D 12. D.

Page 118, Assignment 4.5

1. (i) $-\sin 30°$ (ii) $-\cos 44°$ (iii) $-\tan 50°$
   (iv) $-\tan 60°$ (v) $-\cos 344°$ (vi) $-\sin 60°$
   (vii) $\sin 41°$.
2. (i) $-\sin 30°$ (ii) $-\tan 60°$ (iii) $\cos 39°$
   (iv) $-\cos 51°$ (v) $\tan 60°$ (vi) $-\cos 40°$
   (vii) $\sin 30°$.
3. (i) $\tfrac{40}{41}, \tfrac{9}{40}$ (ii) $-\tfrac{40}{41}, -\tfrac{9}{40}$. 4. $\dfrac{p^2-q^2}{p^2+q^2}, \dfrac{2pq}{p^2-q^2}$.
5. $\tfrac{1}{3}\sqrt{3}$. 6. $\tfrac{4}{5}, -\tfrac{4}{3}$.
9. (i) 2 (ii) $\tfrac{3}{4}$ (iii) $\sqrt{2}+1$.
10. (i) $\{60, 120\}$ (ii) $\{60, 120\}$ (iii) $\{180\}$ (iv) $\{30, 150\}$
    (v) $\{45, 63\cdot4\}$ (vi) $\{0\}$ (vii) $\{0, 70\cdot55\}$
    (viii) $\{0, 60, 120\}$.
11. $\tfrac{5}{13}, -\tfrac{12}{13}$.
12. (i) $\cos A$ (ii) $\tan A°$ (iii) $\sin A°$ (iv) $-\sin A°$
    (v) $-\tan A°$ (vi) $-\cos A°$.
13. (i) $-\dfrac{1}{\sqrt{3}}$ (ii) $-\dfrac{\sqrt{3}}{2}$ (iii) $\dfrac{1}{\sqrt{2}}$ (iv) $\dfrac{\sqrt{3}}{2}$ (v) 1

(vi) $-\dfrac{1}{\sqrt{3}}$ (vii) $-\dfrac{\sqrt{3}}{2}$ (viii) 0 (ix) $-\dfrac{\sqrt{3}}{2}$ (x) $-\dfrac{1}{\sqrt{2}}$.

14. (i) {45, 135} (ii) {120, 240} (iii) {120, 300}
    (iv) {270} (v) {30, 330} (vi) {45, 225}
    (vii) {62·7, 242·7} (viii) {235, 305} (ix) {45, 225}.
15. (i) 23·6, 156·4, 270 (ii) 45, 161·6, 341·6
16. 11·95, 33 m.   17. 625 m.
20. (i) 34°, 146° (ii) 13·5.   21. 1640 m, 502 m.
22. (i) $\dfrac{\pi}{12}, \dfrac{2\pi}{9}, \dfrac{\pi}{3}, \dfrac{7\pi}{6}, \dfrac{3\pi}{2}, \dfrac{5\pi}{2}, 2\pi$
    (ii) 30°, 22·5°, 72°, 135°, 180°, 120°, 150°, 360°.
23. 7·33 cm, 36·6 cm². 24. 770π km. 25. $\tfrac{5}{4}$ rads., 360π cm².
26. (i) sin 2a° cos b° + cos 2a° sin b°
    (ii) cos 2a° cos b° + sin 2a° sin b°
    (iii) sin a° cos b° − cos a° sin b°
    (iv) cos a° cos b° − sin a° sin b°.
27. (i) cos α cos β + sin α sin β
    (ii) sin 2α cos β − cos 2α sin β
    (iii) cos α cos β − sin α sin β
    (iv) cos 2α cos β − sin 2α sin β.
28. (i) $\dfrac{\tan a° - \tan b°}{1 + \tan a° \tan b°}$ (ii) $\dfrac{\tan 2\theta + \tan \alpha}{1 - \tan 2\theta \tan \alpha}$
30. (i) $\dfrac{\sqrt{3}}{2}, \tfrac{1}{2}, \sqrt{3}$ (ii) $\tfrac{24}{25}, -\tfrac{7}{25}, -\tfrac{24}{7}$ (iii) $\tfrac{3}{5}, \tfrac{4}{5}, \tfrac{3}{4}$
32. (i) {60, 169·5, 250·5, 300} (ii) {90, 210, 270, 330}
    (iii) {75·5, 180, 284·5} (iv) {139·1, 319·1}
33. (i) $\{0, \tfrac{2}{3}\pi, \pi, \tfrac{5}{3}\pi\}$ (ii) $\{0, \tfrac{1}{6}\pi, \tfrac{5}{6}\pi, \pi\}$
    (iii) $\{0, \tfrac{1}{3}\pi, \tfrac{2}{3}\pi, \pi, \tfrac{4}{3}\pi, \tfrac{5}{3}\pi\}$
34. (i) 2 sin 44° cos 12° (ii) 2 sin 44° sin 12°
    (iii) 2 cos $\tfrac{1}{2}\pi$ sin $\tfrac{1}{6}\pi$ (iv) 2 cos $\tfrac{1}{2}\pi$ sin $\tfrac{1}{6}\pi$
35. (i) sin 90° + sin 30° (ii) cos 96° + cos 10°
    (iii) $\tfrac{1}{2}$(sin 80° − sin 30°) (iv) $\tfrac{1}{2}$(cos 20° − cos 120°)
36. (i) {0, 90, 120, 240, 270}
    (ii) {45, 135, 210, 225, 315, 300}
    (iii) {45, 60, 135, 225, 300, 315}
37. (i) {53·1, 323·1} (ii) {138·1, 294·9} (iii) {161}
38. (i) tan 2θ, $0 < \theta < \tfrac{1}{4}\pi$ (ii) −1, $\tfrac{1}{4}\pi < \theta < \tfrac{1}{2}\pi$.

CALCULUS

Page 132, Assignment 1.1

1. (i) $\tfrac{5}{3}$ (ii) 6 (iii) −5 (iv) −2 (v) $\dfrac{b}{a}$ (vi) $\dfrac{16-b}{1+a}$ (vii) $\dfrac{k}{h}$.

2. (i) $2x$ (ii) $3x^2$ (iii) $-\dfrac{1}{x^2}$ (iv) $2x+1$ (v) $9x^2$ (vi) $-\dfrac{2}{x^3}$.

3. (i) $2x$ (ii) $6x^2$ (iii) $\dfrac{1}{x^2}$ (iv) $-\dfrac{2}{x^3}$ (v) $2x+1$.
   (vi) $1+\dfrac{1}{x^2}$ (vii) 4 (viii) $1-2x$ (ix) $-\dfrac{3}{x^4}$ (x) $2x$.

4. (i) $8x^7$ (ii) $24x^3$ (iii) 0 (iv) $\dfrac{1}{2\sqrt{x}}$ (v) $2x^{-1/3}$
   (vi) $\tfrac{5}{2}x^{-\frac{1}{2}}$ (vii) $15x^{\frac{1}{2}}$ (viii) $-\tfrac{1}{2}x^{-\frac{3}{2}}$ (ix) $-\dfrac{2}{x^3}$ (x) $-\tfrac{1}{3}x^{-\frac{4}{3}}$.

5. (i) $-\dfrac{25}{x^6}$ (ii) $-\dfrac{1}{2x^3}$ (iii) $\tfrac{1}{2}x^{-\frac{1}{2}} - \tfrac{1}{2}x^{-\frac{3}{2}}$ (iv) $2x+2$
   (v) $12x^3 - x^2 + 1$ (vi) $2x^3 - 3x^2$ (vii) $3x^2 - 5 - \dfrac{1}{x^2}$
   (viii) $2(x+1)$ (ix) $2x + \dfrac{2}{x^2}$ (x) $\tfrac{3}{2}x^{\frac{1}{2}} - x^{-\frac{1}{2}}$ (xi) $2x - \dfrac{2}{x^3}$.

6. (a) $2x - 7 - \dfrac{9}{x^2}$ (b) $3x^{\frac{1}{2}} - \tfrac{7}{2}x^{-\frac{1}{2}} - \tfrac{3}{2}x^{-\frac{3}{2}}$
   (c) $-\tfrac{1}{2}$ (d) 0; $-7\tfrac{1}{2}$.

7. (i) $6(3x+7)$ (ii) $-6(7-3x)$ (iii) $25(5x-4)^4$
   (iv) $3(4+\tfrac{1}{2}x)^5$ (v) $-12(5-2x)^2$ (vi) $\dfrac{-2}{(x-1)^2}$
   (vii) $\dfrac{45}{(5x+2)^4}$ (viii) $pn(px+q)^{n-1}$
   (ix) $-\tfrac{3}{2}(4-3x)^{-\frac{1}{2}}$ (x) $-bp(a-bx)^{p-1}$.

8. $3x^2 + \dfrac{1}{x^2}$.   9. $1 - \dfrac{1}{x^2}$.   10. $2x - \dfrac{2}{x^3}$.

11. (i) $(3-2x)^{-\frac{3}{2}}$ (ii) $\dfrac{3}{2}\left(\dfrac{1}{\sqrt{x}} + \sqrt{x}\right)$ (iii) $3(4x-1)^{-\frac{1}{4}}$.

12. (i) $6x(x^2+1)^2$ (ii) $-x(4-x^2)^{-\frac{1}{2}}$ (iii) $20x(2x^2+1)^4$
    (iv) $\dfrac{3}{2(4-3x)^{\frac{3}{2}}}$ (v) $\dfrac{-1}{(2x+3)^{\frac{3}{2}}}$ (vi) $\dfrac{-2x}{(x^2-4)^2}$.

Page 135, Assignment 1.2

1. (i) -4 (ii) 6 (iii) -1 (iv) 24 (v) $9x^2 - 1$
    (vi) $-\dfrac{1}{a^2}$.

2. (i) 4; $4x - y = 6$ (ii) 12; $y - 12x = 16$
    (iii) 2; $y = 2x - 1$ (iv) $y = 0$ (v) $y = 3$
    (vi) 15; $15x - y = 42$.

3. $4x - y = 7$.   4. $x + y + 3 = 0$.
5. $5x - y = 9$; $A(\tfrac{9}{5}, 0)$, $B(0, -9)$.
6. $(-2, -2\tfrac{1}{2})$.   7. $(1, -1), (-3, 31)$.
8. $y = 2x - 2$; $(-2, -6)$.   9. $y = 5x + 1$, $(2, 11)$.

Page 136, Assignment 1.3

1. A.   2. B.   3. C.   4. A.   5. B.   6. A.
7. A.   8. D.   9. A.   10. E.   11. A.   12. C.

Page 140, Assignment 2.1

1. (i) $x > 2$ (ii) $x < -2$ (iii) $x > 2$ or $x < -2$
    (iv) $-2 < x < 2$.
2. (i) $x < 3$ (ii) $x > -3$ (iii) $x > 3$ or $x < -3$
    (iv) $-3 < x < 3$.
3. $x > 1$ or $x < 0$.   4. $-\tfrac{1}{2} < x < 0$.   5. $1 \leq x \leq 2$.
6. $x > 1\tfrac{1}{2}$ or $x < -2$.   7. $\{x : x \in R, x \neq 0\}$; no value of x.
8. $x > 0$; $x < 0$.   9. $x > 1$; $x < 1$.   10. $x < 1$; $x > 1$.
11. $x > 2$ or $x < -2$; $-2 < x < 2$.   12. $x > 3$; $x < 3$.
13. $x > 2$ or $x < 0$; $0 < x < 2$.
14. $x > 2$ or $x < \tfrac{2}{3}$; $\tfrac{2}{3} < x < 2$.
15. $x > 2$ or $x < 0$; $0 < x < 2$.   16. $\{x : x \in R, x \neq 0\}$.
18. (i) $a > 0$,   (ii) $a < 0$.
19. (i) stationary,   (ii) decreasing,   (iii) increasing.
20. $(0, 2), (1, 3), (2, 2)$.

Page 144, Assignment 2.2

1. $-1, (1, -1)$, minimum.
2. $-6, (1, -6)$, minimum; $21, (-2, 21)$, maximum.
3. $-2, (-1, -2)$, minimum; $2, (1, 2)$, maximum.
4. $3\tfrac{1}{8}, (-\tfrac{3}{4}, 3\tfrac{1}{8})$, maximum.
5. $6\sqrt{3}, (-\sqrt{3}, 6\sqrt{3})$, maximum;
    $-6\sqrt{3}, (\sqrt{3}, -6\sqrt{3})$, minimum.
6. $0, (0, 0)$, point of inflexion on a rising curve.
7. $-27, (-3, -27)$, minimum; $0, (0, 0)$, point of inflexion on a rising curve.
8. $4, (1, 4)$, maximum; $0, (3, 0)$, minimum.
9. $-1, (-1, -1)$, minimum; $-1, (1, -1)$, minimum, $0, (0, 0)$, maximum.
10. $0, (0, 0)$, point of inflexion on a falling curve; $-\tfrac{1}{16}$, $(\tfrac{1}{2}, \tfrac{-1}{16})$, minimum.
11. max. value $1\tfrac{2}{3}$, min. value -9.   12. min value -2.
13. max. value 8.   14. max. value 28, min. value 27.
15. max. value 13, min. value -14.
16. max. value 0, min. value $-\tfrac{4}{27}$.   17. minimum.

Page 144, Assignment 2.3

1. B.   2. A.   3. C.   4. A.   5. D.   6. E.
7. D.   8. B.   9. E.   10. D.   11. B.   12. C.

Page 151, Assignment 3.1

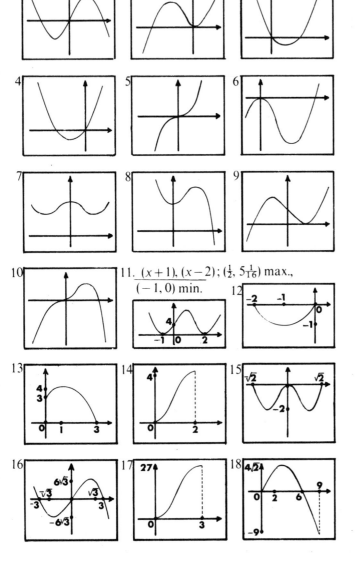

11. $(x+1), (x-2); (\frac{1}{2}, 5\frac{1}{16})$ max., $(-1, 0)$ min.

Page 153, Assignment 3.2

1. 900 m². 3. 31,250 m². 4. 50 cm, 25 cm.
5. $x = 6, h = 3$. 6. 3·25 m. 7. 12.
8. $a = 6, b = 4\frac{1}{2}$ 9. 5. 10. $\frac{R}{3}; \frac{32\pi}{81} R^3$.
11. $x = 15, h = 7\frac{1}{2}; 1687\frac{1}{2}$ cm³.
12. $\frac{12}{5}$, 12 units². 13. 16, 8; 512 cm².

Page 155, Assignment 3.3

1. C. 2. E. 3. C. 4. B. 5. A. 6. E.
7. C. 8. B. 9. E. 10. E. 11. A. 12. C.

Page 160, Assignment 4.1

1. (i) 8 m (ii) 1·6 (iii) 6·4 m.
2. $6b; 6; 30$. 3. $2b^2; 6$. 4. 8 cm² per cm.
5. 36π cm³ per cm. 6. $\frac{l^3}{4}$; 3 cm³ per cm.
7. $1·5\pi$ cm² per sec.
8. 144π cm³ per cm; 576π cm³ per sec.
9. $\frac{1}{4\pi}$ cm per sec; $\frac{1}{2}$ cm per sec. 10. $\frac{1}{6\pi}$ cm per sec.
11. $\frac{2}{3}$ units³ per min; decreasing. 12. $\frac{4}{125\pi}$ m per sec.
13. (i) 125 m (ii) $4\sqrt{10}$ sec (iii) 50 m per sec.
14. (i) 5 m per sec (ii) $101\frac{1}{4}$ m.
15. (i) 21 m per sec (ii) 45 m per sec.

Page 165, Assignment 5.1

1. (i) $\frac{x^4}{4} + C$ (ii) $x^4 + \frac{x^3}{3} + C$ (iii) $\frac{x^3}{3} + \frac{x^2}{2} + x + C$

(iv) x^3+3x^2-2x+C (v) $2x-\dfrac{x^2}{2}+C$

(vi) $3x^5+2x^3+C$.

2. (i) $\dfrac{x^2}{2}+2x^{\frac{1}{2}}+C$ (ii) $\tfrac{2}{3}x^{\frac{3}{2}}+\tfrac{2}{5}x^{\frac{5}{2}}+C$ (iii) $\dfrac{x^4}{4}-\dfrac{1}{x}+C$

(iv) $3x-\tfrac{1}{4}x^2+C$.

3. $\dfrac{(x-2)^3}{3}+C$.

4. $\dfrac{x^3}{3}-9x+C$.

5. $\dfrac{x^3}{3}-2x-\dfrac{1}{x}+C$.

6. $\dfrac{x^3}{3}+\dfrac{x^2}{2}-2x+C$.

7. $\dfrac{x^2}{2}-\dfrac{1}{x}+C$.

8. $\dfrac{x^4}{4}-\dfrac{x^3}{3}-x^2+C$.

9. $\tfrac{2}{3}x^{\frac{3}{2}}+C$.

10. $\tfrac{2}{3}x^{\frac{3}{2}}-2x^{\frac{1}{2}}+C$.

11. $2x^{\frac{1}{2}}-2x^{\frac{3}{2}}+C$.

12. $\dfrac{x^3}{3}-x+\dfrac{3}{x}+C$.

13. $\tfrac{3}{2}x^{\frac{2}{3}}-\tfrac{6}{5}x^{\frac{5}{3}}+\tfrac{3}{8}x^{s}+C$.

14. $-\dfrac{1}{2x^2}-\dfrac{2}{3x^{\frac{3}{2}}}+C$.

15. $\tfrac{1}{2}(16x-12x^2+3x^3)+C$.

16. $\tfrac{3}{8}x^{s}-\tfrac{6}{5}x^{\frac{5}{3}}+\tfrac{3}{2}x^{\frac{2}{3}}+C$.

17. $\dfrac{x^5}{5}+\dfrac{2}{x}-\dfrac{1}{7x^7}+C$.

18. $\tfrac{4}{3}x^{\frac{3}{2}}-\tfrac{8}{7}x^{\frac{7}{8}}+C$.

19. $\dfrac{-1}{6(3x-2)^2}+C$.

20. $4(x-3)^{\frac{1}{2}}+C$.

21. (i) $2x^3+\dfrac{1}{x}-\dfrac{1}{x^2}+C$. (ii) $2x^4-\dfrac{2}{x^2}-3x+C$.

(iii) $\tfrac{3}{5}x^{\frac{5}{3}}-8x^{\frac{1}{2}}+7x+C$.

22. $y=4x^2+\dfrac{4}{x}-4$.

23. $f(x)=x^4+\dfrac{1}{2x^2}-15\tfrac{7}{8}$.

24. $y=\tfrac{2}{3}x^{\frac{3}{2}}-2x^{\frac{1}{2}}$.

25. $f(x)=\tfrac{3}{5}x^{\frac{5}{3}}+\tfrac{3}{2}x^{\frac{2}{3}}+\tfrac{74}{5}$.

26. (i) $\dfrac{(2t+1)^8}{16}$ (ii) $(3x-4)^{\frac{1}{3}}$ (iii) $(2x-1)^{\frac{1}{2}}$

(iv) $-\dfrac{(3-4x)^6}{24}$ (v) $-\tfrac{1}{2}(1-4x)^{\frac{1}{2}}$ (vi) $8(\tfrac{1}{2}t-4)^{\frac{1}{4}}$.

Page 166, Assignment 5.2

1. B. 2. C. 3. C. 4. A. 5. E. 6. D.
7. E. 8. B. 9. D. 10. A. 11. B. 12. A.

Page 170, Assignment 6.1

1. $\tfrac{1}{2}$. 2. $\tfrac{2}{3}$. 3. $1\tfrac{1}{2}$. 4. 0.
5. $44\tfrac{1}{4}$. 6. $1\tfrac{1}{2}$. 7. $\tfrac{14}{3}$. 8. -4.
9. $\tfrac{14}{3}$. 10. $\tfrac{8}{3}$. 11. 10. 12. $k^3+\dfrac{1}{k}-2$.
13. 2. 14. $5\tfrac{1}{3}$. 15. $\tfrac{20}{3}$. 17. 8.
18. 16.

Page 172, Assignment 6.2

1. $\tfrac{16}{3}$. 2. 6. 3. 16. 4. 32.
5. $\tfrac{14}{3}$. 6. $\tfrac{14}{3}$. 7. (i) $\tfrac{20}{3}$, (ii) $\tfrac{4}{3}$.
8. $\tfrac{1}{2}$. 9. $\dfrac{9\pi}{2}$. 10. π. 12. 8; $-\tfrac{16}{3}$.
13. $16\tfrac{1}{2}$. 15. $3\tfrac{1}{3}$.

Page 174, Assignment 6.3

1. $\tfrac{4}{3}$. 2. $\tfrac{32}{3}$. 3. $\tfrac{1}{6}$. 4. $\tfrac{4}{3}$.
5. $\tfrac{4}{3}$. 6. 8. 7. $\tfrac{4}{3}$. 8. $\tfrac{8}{3}$.
9. $\tfrac{9}{8}$. 10. 9. 12. $\tfrac{128}{15}$. 13. $10\tfrac{2}{3}$.

Page 177, Assignment 6.4

1. $\dfrac{9\pi}{5}$. 2. $\dfrac{21\pi}{5}$. 3. 12π. 4. $\dfrac{5\pi}{6}$.

5. 28π. 6. $\dfrac{\pi}{4}$. 7. $\dfrac{\pi}{5}$. 8. $\dfrac{4\pi}{3}$.

9. $\dfrac{441\pi}{10}$. 10. $\dfrac{10\pi}{3}$.

Page 178, Assignment 6.5

1. $\tfrac{4}{3}, \pi$. 2. $\tfrac{16}{3}, \dfrac{64\pi}{5}$. 3. $\dfrac{16\sqrt{2}}{15}, \dfrac{4\pi}{3}$.

4. $1, \dfrac{7\pi}{3}$. 5. $\tfrac{32}{3}, \dfrac{512\pi}{15}$.

6. $A(1, 2), B(1, -2), \dfrac{112\pi}{15}$.

7. (i) 64π, (ii) $\dfrac{384\pi}{7}$.

8. (i) $\dfrac{32\pi}{3}$, (ii) $\dfrac{128\pi}{15}$.

9. $V_1 = \dfrac{32\pi}{3}$, $V_2 = \pi$. 10. 16π; No, 24π.

Page 180, Assignment 6.6

1. C. 2. D. 3. D. 4. D. 5. A. 6. C.
7. D. 8. C. 9. B. 10. E. 11. B. 12. C.

Page 184, Assignment 7.1

1. $-6 \sin x$. 2. $-3 \cos x$. 3. $\cos x + 2 \sin x$.
4. $-4 \sin x - 3 \cos x$. 5. $2x + \cos x$.
6. $\tfrac{1}{2} \sin x$. 7. $3 \cos 3x$. 8. $-4 \sin 4x$.
9. $3 \cos(3x+2)$. 10. $-4 \sin(4x-5)$.

11. $3 \sin(2-3x)$. 12. $-\cos(1-x)$. 13. $4 \sin^3 x \cdot \cos x$.
14. $-5 \cos^4 x \cdot \sin x$. 15. $-2(2-\sin x) \cos x$.
16. $2 \cos 2x - 4 \sin 4x$. 17. $-\sin x - 2 \cos x \cdot \sin x$.
18. $\cos x - 2 \cos x \cdot \sin x$. 19. $-2(1+\cos x) \sin x$.
20. $3 \cos 3x + 4 \cos^3 x \cdot \sin x$.
22. (i) $y + 2x = \pi$ (ii) $4y + 4x = 2 + \pi$ (iii) $y + x = \pi - 1$.
24. $90 < x \leqq 180$. 25. $0 \leqq x < 45$.
26. $0, 180, 360$ give maximum stationary values.
$90, 270$ give minimum stationary values.
27. $45, 225$ give maximum stationary values.
$135, 315$ give minimum stationary values.

28. $\left(\dfrac{3\pi}{4}, \sqrt{2}\right)$ max. turning point.

$\left(\dfrac{7\pi}{4}, -\sqrt{2}\right)$ min. turning point.

29. $x = \dfrac{\pi}{6}$ gives a max. stationary value of $\dfrac{3\sqrt{3}}{4}$.

$x = \dfrac{5\pi}{6}$ gives a min. stationary value of $\dfrac{-3\sqrt{3}}{4}$.

$x = \dfrac{3\pi}{2}$ gives a P.I. on a rising curve.

Page 186, Assignment 7.2

1. $\tfrac{1}{2} \sin 2x + C$. 2. $-\tfrac{1}{5} \cos 5x + C$.
3. $-\tfrac{1}{3}(3x+2) + C$. 4. $\tfrac{1}{4} \sin(4x+1) + C$.
5. $-\sin(-x+2) + C$. 6. $\tfrac{1}{3} \cos(2-3x) + C$.
7. $-\tfrac{1}{6} \sin 6x + C$. 8. $\tfrac{1}{4} \cos 4x + C$.
9. $-\sin(x+1) + C$. 10. $\tfrac{1}{2} \cos(4x+2) + C$.
11. $\dfrac{1}{2}\left(x + \dfrac{\sin 2x}{2}\right) + C$. 12. $\dfrac{1}{2}\left(x - \dfrac{\sin 2x}{2}\right) + C$.
13. 1. 14. 2.

15. 0.   16. $\tfrac{1}{3}$.   17. $-\tfrac{1}{3}$.

18. (i) 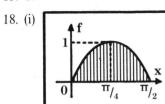   (ii)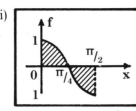

Area = 1 unit^2 Area = 1 unit^2

19. $1 + \sin 2\theta$; $1 + \dfrac{\pi}{2}$.   20. $\dfrac{\pi}{4}$, 1 unit^2.   21. $\dfrac{\pi^2}{8}$ units^2.

22.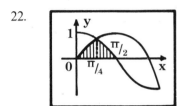

Area = $2 - \sqrt{2}$ units^2

Volume = $\dfrac{\pi^2}{4} - \dfrac{\pi}{2}$ units^3

23. $1\tfrac{1}{4}$ unit^2.

Page 188, Assignment 7.3

1. C.   2. B.   3. D.   4. C.   5. E.   6. E.
7. D.   8. B.   9. E.   10. E.   11. B.   12. A.

Page 190, Assignment 7.4

1. (i) $4x$ (ii) $-\dfrac{3}{x^4}$ (iii) $-\dfrac{4}{x^3}$.

2. $\tfrac{3}{2}x^{\tfrac{1}{2}} + \tfrac{3}{2}x^{-\tfrac{1}{2}} - \dfrac{4}{x^3}$.   3. $3x^2(2-2x^3)^{-\tfrac{3}{2}}$.

4. $\tfrac{3}{2}\sqrt{x} + \dfrac{1}{\sqrt{x}}$.   5. $\tfrac{3}{2}x^{\tfrac{1}{2}} + x^{-\tfrac{1}{2}} + \tfrac{7}{2}x^{-\tfrac{3}{2}}$.

6. (i) $-\dfrac{4}{x^5}$ (ii) $3x^2 - 7 - \dfrac{1}{3x^2}$.   7. $-\tfrac{4}{15}$.

8. (i) $1 + 12t - 9t^2$ (ii) $-\dfrac{2}{t^3} - \dfrac{3}{t^4}$ (iii) $8t - \dfrac{2}{t^3}$

(iv) $k\left(t^{k-1} + \dfrac{1}{k^{t+1}}\right)$.

9. (i) $\tfrac{1}{6}$ (ii) $0, \tfrac{3}{2}$.   10. $\pm\tfrac{1}{2}$.

11. (i) $-\tfrac{1}{2}(x^{-1\tfrac{1}{2}} + 5x^{1\tfrac{1}{2}})$ (ii) $1 - 2\cos 2x$ (iii) 0.

12. $-\dfrac{5\sqrt{2}}{2}$.   13. $y = x + 8$; $(\tfrac{2}{3}, -\tfrac{22}{27})$.

14. $y + 2x + 4 = 0$, $(-1, 6)$.   15. $7, -7$; $y + 5x = 2$.

16. $(\tfrac{1}{2}, 2\tfrac{1}{2}), (-\tfrac{1}{2}, -2\tfrac{1}{2})$.   17. $(-1, 1\tfrac{1}{2})$.

18. $-1 < x < 7$; $-1 < x < 3$.

20. $\tfrac{1}{3} < x < 1$.   21. $\dfrac{\pi}{3} < t < \pi$.

22. Max. $(-2, 16)$, $(\tfrac{1}{2}, \tfrac{3}{8})$, Min. $(0, 0)$; $(0, 0)$, $(-1 \pm \sqrt{3}, 0)$.

23. Max. $(1, 4)$, Min. $(3, 0)$.   24. 18π.

25. Max. $\tfrac{3}{4}$, Min. $\tfrac{1}{2}$.   27. $r\sqrt{2}, \tfrac{1}{2}r\sqrt{2}$; r^2.

28. $6 + \sqrt{3}, \tfrac{3}{2}(5 - \sqrt{3})$ m.   29. $16 \text{ cm}^2/\text{sec}$.

30. $\dfrac{44}{9\pi}$ cm/min.   31. $-1, 3$.

32. $y = \dfrac{5x^2}{2} + \dfrac{1}{x^2} - \dfrac{1}{4}$.   33. $y = -\dfrac{6}{x} - 4x^2 + 19$.

34. (i) $4x + \dfrac{5}{x} + 2x^{\tfrac{3}{2}} + C$ (ii) $x^4 + x + \dfrac{3}{2x^2} + C$.

35. $\dfrac{x^3}{3} - \dfrac{1}{x} + C$.   36. $\dfrac{x^3}{3} + \dfrac{3x^2}{2} - 5x + C$.

37. $-\dfrac{2}{x} - \dfrac{1}{2x^2} + C$.   38. $-\dfrac{1}{x} - \dfrac{2}{\sqrt{x}} + C$.

39. $-\tfrac{2}{3}(3-x)^{\tfrac{3}{2}} + C$.   40. $-\tfrac{1}{4}(4-3t)^{1\tfrac{1}{3}} + C$.

41. $\tfrac{1}{12}(2+3t)^4 + C$.   42. $-\dfrac{\cos 2x}{4} + C$.

43. $f(x) = \frac{2}{3}x^3 - 18x - 20$.   44. $y = \dfrac{x^4}{4} - 2x^3 + \frac{9}{2}x^2 - \frac{3}{4}$.

45. $\frac{3}{4}$.   46. $13\frac{1}{3}$.   47. 6.   48. $\frac{1}{2}$.   49. π.   50. 1.

51. $227\frac{5}{9}$.   52. (i) 128 units^2 (ii) 4.

53. $5\frac{1}{4}\pi \text{ units}^3$.   54. $207\pi \text{ units}^3$.

55. (i) $\sqrt{2}$ (ii) 1.   56. (ii) $\dfrac{\pi}{3}, \dfrac{2\pi}{3}$.